Annual Meeting of the American Electrophoresis Society 2017 (AES)

AA009005

Topical Conference at the 2017 AIChE Annual Meeting

Minneapolis, Minnesota, USA
29 October - 3 November 2017

ISBN: 978-1-5108-5775-9

Printed from e-media with permission by:

Curran Associates, Inc.
57 Morehouse Lane
Red Hook, NY 12571

Some format issues inherent in the e-media version may also appear in this print version.

Printed by Curran Associates, Inc. (2018)

For permission requests, please contact AIChE
at the address below.

AIChE
120 Wall Street, FL 23
New York, NY 10005-4020

Phone: (800) 242-4363
Fax: (203) 775-5177

www.aiche.org

Additional copies of this publication are available from:

Curran Associates, Inc.
57 Morehouse Lane
Red Hook, NY 12571 USA
Phone: 845-758-0400
Fax: 845-758-2633
Email: curran@proceedings.com
Web: www.proceedings.com

TABLE OF CONTENTS

Some abstracts in this book refer to images that were not made available by the publisher

503746 Organelle Separation with a Microfluidic Ratchet

Abstract:

Heterogeneity in size and function of intracellular organelles is intimately associated with dynamic cellular processes. Such processes include fusion, fission, biomolecular synthesis within the organelle, biomolecular transport, and degradation. Organelle morphology relates to the dysfunction of cellular processes such as organelle quality control, oxidative stress, decreased respiratory chain function in mitochondria, lipid storage, and programmed cell death, which are associated with diseases including Alzheimerâ€™s disease, cardiovascular disease, diabetes, cancer, and aging among others. Organelle subpopulations thus play critical roles in the onset of disease and their recovery, extraction and purification from cells and tissue is imperative to unravel the underlying biological mechanisms related to size and function. Current methods to separate organelles can however not access subpopulations by size and in most cases, organelle recovery is scarce and impurities are common. Here, we present a novel ratchet-based separation method designed for the fractionation of sub-μm sized organelles. The migration mechanism requires tailored microenvironments as well as subtle tuning of electrokinetic and dielectrophoretic driving forces. To characterize the ratchet migration mechanism, we have thus developed a numerical model allowing to predict the regime of driving parameters under which it occurs. We demonstrate that the numerical model can be applied to hundreds of particles, allowing the study of the separation performance. Moreover, micrometer and sub-micrometer particles can be separated with baseline resolution in less than one minute. We further show that experimental observations match the numerical model in very good agreement for polystyrene particles and that mitochondria isolated from rat liver follow the predicted migration mechanism.

503739 Cell Surface Complexity Modulates Membrane Capacitance and Differentiation of Human Neural Stem Cells

Abstract:

Human neural stem cells have potential as therapeutic candidates for the treatment of a number of neurodegenerative diseases or neurotraumas. Important issues that need to be addressed to improve the utility of these cells in clinical trials include characterization of the heterogeneity in stem cell populations and ways to modulate cell differentiation. Our previous studies have shown differential membrane capacitance values for cells with astrogenic and neurogenic fate. In the present study we have investigated the modulatory effect of cell surface complexity on membrane capacitance and differentiation of these human neural stem cells.

Membrane capacitance was measured using the 3DEP analyzer, a microdevice that uses electric fields to quickly quantitate capacitance. The cell surface of human neural stem cells was altered by treating cells with N-acetylglucosamine (GlcNAc), which feeds into N-glycosylation pathways leading to the formation of complex, highly-branched sugars. Treatment of human neural stem cells with GlcNAc significantly increases complex branching on the cell surface (lectin LPHA, untreated 18485+/-2069 SEM mean fluorescence intensity (MFI), GlcNAc treated 40863 +/- 4428 SEM. MFI, $p<0.001$). Increasing complex branching on human neural stem cells significantly increases membrane capacitance (untreated 7.5 +/- 0.18 s.e.m. mF/m^2, GlcNAc treated 9.3 +/- 0.47 s.e.m. mF/m^2; $p<0.001$)), showing that cell surface glycosylation patterns impact membrane capacitance values. Modulation of cell surface complexity also affects cell differentiation. Increase in cell surface N-glycosylation increases the percentage of cells expressing undifferentiated cell markers after the cells were in differentiation conditions. Taken together these results indicate a direct association between membrane capacitance, cell surface complexity and differentiation of human neural stem cells.

503720 Characterizing Human Embryonic Stem Cells Function with Dielectrophoresis and Flow Cytometry

Abstract:

Characterizing Human Embryonic Stem Cells Function With Dielectrophoresis And Flow Cytometry

Tayloria Adams, Clarissa Ro, Shubha Tiwari, and Lisa Flanagan

Neurology Department, University of California, Irvine

Human embryonic stem cells (hESCs) provide great opportunities in stem cell therapeutics because they can differentiate into the three germ layers: ectoderm, mesoderm, and endoderm. More specifically, hESCs can be directed toward human neural stem/progenitor cell (hNSPC) differentiation and used to treat neurological diseases and injuries. Sufficiently characterizing hNSPCÕs functional behavior before using them in transplant therapy is essential to the development of reliable therapeutic treatment options. Having measures that will accurately reflect cell phenotype after transplantation is critical. Currently, the process for cell characterization is challenging due to a lack of biomarkers that provide an adequate picture of hNSPCs specific functions. Therefore, weÕve implemented dielectrophoresis (DEP), a label-free characterization technique that uses nonuniform electric fields, to determine cellsÕ dielectric properties such as membrane capacitance (C_{mem}). Additionally, hNSPCs were analyzed by flow cytometry to characterize surface protein expression answering the scientific question Òare there correlations between C_{mem}, measured by DEP, and protein expression?Ó

Three sets of hESC-derived hNSPCs (Shef4-1, Shef4-2, Shef4-3) were derived from passage 6 EZ spheres and established as monolayers before analysis. The Shef4-1, Shef4-2, and Shef4-3 nomenclature indicates that these cells were derived from the same set of EZ spheres but separately differentiated to form hNSPCs, as would be done to generate sufficient numbers of stem cells for therapeutic purposes. Once established as monolayer hNSPC cultures, their cell size, DEP spectra, C_{mem}, and cell surface protein expression were quantified at a variety of passage numbers to determine lot-to-lot variability or variability over passaging.

Results show that cell diameter varies across lots with Shef4-1 > Shef4-2 > Shef4-3. Similarly, there is variability across lots in the DEP spectra, which is corroborated with C_{mem}. At early and late passages between 0-100kHz the DEP spectra is shifted right for Shef4-2 as compared to Shef4-1 and Shef4-3 (Figure 1A and 1B). The corresponding C_{mem} at early and late passages, Figure 1C and 1D, shows that each cell batch is different with statistical significance (one-way ANOVA). Across passages, Shef4-1 and Shef4-2 C_{mem} gradually decreases and Shef4-3 C_{mem} fluctuates minimally. Alpha6 and alphaV integrin markers were assessed by flow cytometry for each batch of cells, and the % protein expression was plotted with C_{mem}, Figure 1E and 1F. For alpha6, there is a strong correlation to Shef4-2 C_{mem} (R^2=0.882), and essentially no correlation to Shef4-3 (R^2=0.081). For alphaV, there is a weak correlation for Shef4-2

(R^2=0.244), and no correlation for Shef4-3 (R^2=0.082). Shef4-1 was not included because alpha6 and alphaV expression was low (less than 14%).

Figure 1. hESC derived hNSPCs were characterized using DEP and flow cytometry. Shef4-1, Shef4-2, and Shef4-3 DEP spectra at **(A)** early (p4) and **(B)** late (p6 and p7) passage. Average C_{mem} for **(C)** early and **(D)** late passage Shef4-1, -2, and -3 cells. **(E)** Alpha6 and **(F)** alphaV integrin expression with the associating C_{mem} (passages are matched). One-way ANOVA posthoc Tukey test conducted for multiple comparisons, **p <0.01 and ****p < 0.0001.

Significant lot-to-lot variability exists among hESC-derived hNSPCs, and DEP plus flow cytometry provide a good quantitative assessment of cell phenotype variability. Shef4-2 C_{mem} correlates well to alpha6 while no correlations were seen with alphaV. Alpha6 and alphaV integrins are important for cell migration and differentiation. Linking integrin expression with DEP measurements provides additional insights to the biological meaning of C_{mem} to expand beyond the core definition, which is the ability to store charge. Effectively discerning the functional behavior of stem cells before their use in transplants is essential to advance stem cell therapeutics.

501572 Electrical Detection of Zika Virus on Paper Microchip with Silver-Graphene-Nano-Composite Electrode

Abstract:

Electrical detection of Zika virus on paper microchip with silver-graphene-nano-composite electrode

Mohamed Shehata Draz, Harini Lakshminarayanan, Manasa Venkataramani, Kamyar Mehrabi, Maryam Moazeni, Hadi Shafiee

Division of Engineering in Medicine, Department of Medicine, Brigham and Womenâ€™s Hospital, Harvard Medical School, Boston, MA 02115, USA

Abstract:

Zika virus is a member of the flavivirdae family. It is primarily transmitted by *Aedes* mosquito to human. Recently, there are increasing reports confirming that Zika can pass from mother to fetus during pregnancy causing severe microcephaly and other birth defects. There are currently no effective vaccines for Zika infection and the laboratory-based methods for Zika detection are challenging. There are no specific clinical features of Zika infection and the developed symptoms can be mistaken for those of dengue fever and Chikungunya infection. The current Zika infection detection methods mainly rely on identifying either viral RNA or IgM antibodies using RNA nucleic acid Testing (RNA NAT) or MAC-ELISA assay, respectively. However, RNA is detectable in whole blood for only 14 days after the onset of symptoms, and thus a negative RNA NAT result does not exclude infection. On the other hand, the results of Zika MAC-ELISA are unreliable due to the cross-reactivity with other Flaviviruses. The development of rapid, low-cost Zika infection diagnostics can shift the paradigm in the management of infection. Here, we developed a paper-based microchip with printed microelectrodes for Zika virus detection using electrical sensing of viral lysate. The microchip was fabricated using cellulose paper substrate (0.18 mm in thickness), a thin plastic sheet (0.1 mm in thickness), and a double-sided adhesive (DSA). Finger-like integrated electrodes were screen-printed on the paper-plastic substrate using silver-graphene nanocomposite followed by 120 min drying step at 50 ^{0}C. The target viruses were captured on magnetic beads conjugated with anti-Zika antibodies. The captured viruses were then washed with a low-electrically conductive solution to remove the electrically conductive background solution. The captured viruses were then lysed using 1% Triton X-100 to release the charged molecules into the non-ionic background solution. This viral lysate step changes the electrical conductivity of the solution that can be detected as a biomarker signal using impedance spectroscopy on-chip. The results showed that the impedance magnitude of the bulk samples were directly proportional to the concentration of the virus in spiked samples. Our results showed that the chip was able to detect Zika viruses in samples with viral loads as low as 10^2 copies/Î¼L. To evaluate the specificity of the microchip, we challenged the system with Zika virus and different non-targeted viruses, such as Dengue-1, Dengue-2, HSV, HIV, and CMV. The results confirmed the ability of the microchip to specifically detect Zika virus from other non-targeted viruses including closely related flaviviruses such as Dengue-1 and Dengue-2. The developed assay is simple, rapid,

cost-effective and sensitive and it can potentially be used for point-of-care diagnosis of Zika virus.

497609 Characterization of Â Dielectrophoretic Response of Candida Cells Using 3D Carbon-Electrode Dielectrophoresis

Abstract:

Here we present preliminary results towards the characterization of the dielectrophoretic spectrum of different *Candida* cells. In the last two decades, *Candida* infection has emerged as the fourth most common cause for morbidity and mortality in hospitalized patients in the United States [1]. Five *Candida* species have been found to be responsible for ninety percent of the all the cases of *Candida* infection: *Candida albicans*, *Candida glabrata*, *Candida tropicalis*, *Candida krusei*, and *Candida parapsilosis* [2]. The current diagnosis of Candida infection begins with the culturing of tens of milliliters of blood from the patient, which usually takes several days for detection. During the time waiting for culture results, infection may become progressively worse, leading to complications in treatment or death. These complications can be exacerbated by beginning treatment regimens that may not be effective for the particular *Candida* species causing the infection. Hence, there is a need for rapid, species-specific detection of infection from *Candida* cells. Dielectrophoresis (DEP) is a promising technique for the isolation and detection of the *Candida* cells within a few hours. Therefore, it is important to characterize the DEP spectrum of these cells to understand how DEP parameters (i.e. frequency, voltage, signal waveform) may contribute to the degree to which cells can be trapped and separated for detection. To date, we have characterized the frequency dependent behavior of *Candida albicans*, *Candida tropicalis*, and *Candida parapsilosis*.

Strains of C. albicans, C. tropicalis, and C. parapsilosis were cultured in Yeast Malt Broth (YMB) supplemented with a 40% glucose solution at a 1:100 glucose to YMB ratio. Citric Acid was added to the YMB media to adjust the pH to 3.5. Cells were grown in suspension and dynamic incubation (shaking at 215 rpm) at 28°C for two days. For DEP experiments, the experimental media was prepared for cell suspension and buffer solution by dissolving 15% sucrose, 0.3% dextrose, and 0.1% bovine serum albumin (BSA) in DI water. The media conductivity was 12.62 μS/cm. We prepared the experimental cell suspension by suspending 300 μl of the cell culture in 5 ml of the experimental media followed by washing and re-suspending in experimental media. The cell concentrations of the cell suspensions were in the order of 10^7 cells/ml. The fabrication of the carbon-electrode DEP device has been detailed by our group in previous publications[3]. We flowed 58 μl of cell suspension through the electric field of a sinusoidal signal of constant 20 Vpp magnitude and varying frequencies ranging from 10 kHz to 1 MHz. 100 μl of clean buffer was flowed through the DEP chip at a constant flow rate of 10 μl/min to wash any trapped cells. We recorded the release of the cells at the end of the electrode arrays upon turning off the signal. Subsequent image analysis was completed with ImageJ, an image-processing software, to measure the average intensity throughout the cell elution. The results from the image analysis (Figure 1a) show the positive DEP response of cells with different frequency. Figure 1b, 1c and 1d shows the *C. albicans*, *C. tropicalis* and *C. parapsilosis* cells trapped on the carbon electrodes under positive DEP, respectively.

Ongoing work includes characterizing the DEP spectrum of *C. glabrata* and *C. krusei*. We intend to find their cell-wall capacitance from these DEP spectra so that they can be used in any numerical study involved the *Candida* cells.

Future work includes selectively isolating and concentrating any *Candida* species from other *Candida* species using DEP.

References:

ADDIN Mendeley Bibliography CSL_BIBLIOGRAPHY [1] D.R. Andes, N. Safdar, J.W. Baddley, G. Playford, A.C. Reboli, J.H. Rex, J.D. Sobel, P.G. Pappas, B.J. Kullberg, Impact of treatment strategy on outcomes in patients with candidemia and other forms of invasive candidiasis: A patient-level quantitative review of randomized trials, Clin. Infect. Dis. 54 (2012) 1110–1122. doi:10.1093/cid/cis021.

[2] S. Chakravarthi, N. Haleagrahara, A comprehensive review of the occurance and management of systematic candidiasis as an opportunistic infection, Microbiol. J. 1 (2011) 1–7. doi:10.3923/mj.2011.1.7.

[3] M. Islam, R. Natu, M.F. Larraga-Martinez, R. Martinez-Duarte, Enrichment of diluted cell populations from large sample volumes using 3D carbon-electrode dielectrophoresis, Biomicrofluidics. 10 (2016) 033107. doi:10.1063/1.4954310.

499242 A Method for the Sustainable Synthesis of Carbon Fibers Using Dielectrophoresis of Bacteria and Pyrolysis

Abstract:

A Method for the Sustainable Synthesis of Carbon Fibers using Dielectrophoresis of Bacteria and Pyrolysis

Devin Keck, Monsur Islam and Rodrigo Martinez-Duarte

Mechanical Engineering Department, Clemson University

Clemson, SC, USA

Presented here are preliminary experiments facilitating the use of the cellulose-extruding bacteria *Gluconoacetobacter xylinus* as a sustainable tool for the production of carbon fibers. *G. xylinus* has the inherent ability to convert different kinds of sugars into highly crystalline cellulose nanofibrils. The random movement of the bacteria within the cellulose culture leads to a porous cellulose scaffold with no apparent order. Additionally, bacterial cellulose has higher purity and exhibits superior mechanical properties when compared to the cellulose extracted from plants[1].

Results presented in previous years focused on how the heat treatment of cellulose scaffolds functionalized with metal precursors yielded a porous tungsten carbide. Currently, we are focusing on the positioning of *G. xylinus* cells using Dielectrophoresis (DEP) to grow cellulose fibers from specific locations, which will then be carbonized to create individual carbon fibers. DEP refers to the movement of electrically polarized cells in response to a non-uniform electric field of specific frequency and magnitude. Specifically, positive DEP will be used here to position the *G. xylinus* cells onto planar electrodes. The bacteria will be perfused in a continuous flow of nutrient-rich media, allowing for the continuous extrusion of straight fibers. The ultimate goal is to develop a scalable platform for the continuous production of cellulose nanofibers that can then be converted to carbon. This would become a more sustainable alternative to current practices of cellulose production by eliminating the cost and energy required for forest management, cellulose extraction, and cellulose purification. Most importantly, it eliminates the use of oil in the production of carbon fibers.

In this work, we present the initial characterization of the response of *G. xylinus* to electrical stimuli of varying frequencies as well as the protocol developed to convert bacterial cellulose into carbon. The dielectrophoretic response was characterized using AC signals of 1, 5, 10, 15, and 20MHz frequencies at a constant voltage of 20 V_{pp}. The results indicate a positive DEP response across the spectrum of 1-20MHz with the highest trapping at a frequency in the range

of 15-20 MHz (figure 1). Additional work focused on the carbonization of bacterial cellulose nanofibers. Amorphous carbon was successfully achieved (figure 3) through heat treatment of bacterial cellulose to 900 C in a nitrogen atmosphere following a 5 C/min heating ramp. A dwell time of 30 minutes at a temperature of 300 C was included in the heat treatment to facilitate the release of oxygen from the sample to prevent burning at higher temperatures. Such step is also necessary because the removal of the hydroxyl groups stabilizes the carbon polymer by producing conjugated double bonds, forming an aromatic structure [2]. An optimal heating rate was determined to be 5 C/min to achieve high carbon yield without making the process prohibitively long. Although we observed cellulose to carbonize at around 400 C, 900 C was used as the maximum temperature to ensure a high degree of carbonization [3].

Ongoing work is on characterizing the fiber properties and dimensions depending on the suspending media and its flow rate over the cells. The goal is to understand how the conditions in the microfluidic platform lead to the tailoring of the fiber properties and a maximized throughput.

Figure 1: Characterizing the response of G. xylinus to Carbon DEP

Figure 2: Carbonized Cellulose Nanofibers

References

1. Huang, Y., Zhu, C., Yang, J., Nie, Y., Chen, C., & Sun, D. (2014). Recent advances in bacterial cellulose. *Cellulose*, *21*(1), 1-30.

2. McGrath, T. E., Chan, W. G., & Hajaligol, M. R. (2003). Low temperature mechanism for the formation of polycyclic aromatic hydrocarbons from the pyrolysis of cellulose. *Journal of Analytical and Applied Pyrolysis*, *66*(1), 51-70.

3. Rhim, Y. R., Zhang, D., Rooney, M., Nagle, D. C., Fairbrother, D. H., Herman, C., & Drewry, D. G. (2010). Changes in the thermophysical properties of microcrystalline cellulose as function of carbonization temperature. *Carbon*, *48*(1), 31-40.

501366 Microscale Extraction of Rare Earth Elements Using Biosorption and Dielectrophoresis

Abstract:

Rare earth elements (REEs) are a group of chemically similar elements used in catalytic converters, aircraft engines, hi-tech products, and many other applications for which there is no equal substitute. Despite their name, most are abundant in nature but are challenging to extract. Various methods including salmon milt, fractional step, ion exchange and solvent extraction have been utilized in extracting REEs from the crust but the economics of the processes is burdensome. Apart from the economic considerations, other setbacks such as the disruption of the ecosystem, environmental degradation also call for an alternative extraction methodology. In this work, we present an alternative, cost-effective and environmentally friendly method of extracting REEs from the crust. The method is a synergy of biosorption and dielectrophoresis: Gram-negative bacteria-Cupriavidus necator is grown and used to extract the REEs and dielectrophoresis with subsequent cellular digestion is used to recover the REEs.

501485 Electrophysiology of Borrelia Burgdorferi

Abstract:

Laboratory tests for active and previous infections with *Borrelia burgdorferi* have been known to follow a two-test approach of combining enzyme immunoassay(EIA) or immunofluorecesnt assay (IFA) with Western immunoblot. Advances in disease detection techniques have also introduced Polymerase Chain Reaction (PCR) with known limitations. Apart from the fact that culturing *Borrelia burgdorferi* may require weeks, immunoassays for antigen detection are not as sensitive as expected. Since DNA testing could not differentiate live and dead cells and the possibility of laboratory contamination with amplified DNA, which poses a risk of false-positive results, is almost inevitable, PCR does not prove useful for the detection of Lyme disease. Dielectrophoresis, a method that manipulates cells in non-uniform electric field, has been proposed here as a potential substitute for serology and PCR. This present work seeks to obtain the electrical properties of *Borrelia burgdorferi* by fitting cross-over-conductivity data using a three-shell cell model.

495759 Understanding the Role of Electrokinetics-Hydrodynamics in Electrical-Field Based Soil Remediation

Abstract:

The applied electrical field mediated process of cleaning a contaminated soil matrix in order to remove the contaminants (i.e. organic materials, inorganic and heavy metal ions, etc.) is affected by the electrokinetic forces. These forces help to move the contaminants from one position of the soil to another where they can be drained. The technique is very promising and it can be made more efficient through better understanding of the fundamental principles that govern this process. A useful way to achieve this aim is through mathematical modelling of the fluid flow and the solutesâ€™ transport of these contaminants. In this work, we present an analysis of the role of electrokinetics-hydrodynamics in facilitating soil cleaning. In particular, we assume a soil matrix with rectangular capillaries and a Newtonian buffer to compute the electroosmotic flow. The electrokinetic force is computed by using the Poisson-Boltzmann equation coupled with the Debye-Huckel approximation assuming a uniform charge on the capillary walls. The macro-heterogeneity of the capillary in the soil matrix (i.e. orientation of the capillary) is investigated along with other relevant parameters. The results are of importance in order to understand contaminant transport under the presence of capillary-wall reactions and in order to improve the efficiency of the process.

503719 Electrohydrodynamics of a Viscous Drop

Abstract:

Electrohydrodynamics (EHD) of a viscous drop is of great relevance in engineering and industrial applications. In this talk we present both modeling and numerical results from investigating the electrohydrodynamics of a viscous drop under an electric field. We will illustrate how surfactant (surface active agent) may lead to different drop EHD, both dielectric drop or conducting drop. Numerical algorithms are developed to efficiently simulate the drop EHD, for a range of electric field strength. Under an electric field of moderate strength the steady drop shape compares well with the numerical simulation results, both in good agreement with experimental results. Under a strong electric field, several drop EHD is observed depending on the combination of physical parameters. Furthermore inertia effects are found to be important when the drop undergoes extreme deformation under a strong electric field. Finally we extend our EHD model to include electrokinetic effects and illustrate how EHD is affected by key electrokinetic phenomena.

503695 Electro-Hydrodynamic Behavior of Soft Liquid Metal Plugs Under Low Voltages

Abstract:

This work characterizes the behavior of a eutectic alloy of gallium and indium (75% Ga, 25% In, by weight, 'EGaIn') in response to electric fields. The metal is a liquid at room temperature (M.P., 15.5 °C) and exhibits low toxicity. These fluidic metals may be injected into microfluidic systems, fibers, and capillary networks to form soft electronic devices that are soft and compliant. Once injected, the metal remains in its place because of the adhesive nature of its thin native oxide. Preventing the oxide adhesion within microchannels enables reversible actuation of EGaIn. Actuating liquid metals may be useful for soft actuators, reconfigurable optical displays, frequency tunable antennas, and other opto-fluidic technologies.

In this work, we utilize low voltages (<2 V) to reversibly move droplets of EGaIn through microchannels. Pre-wetting the channels with an aqueous solution prior to injecting the metal prevents oxide adhesion; the water forms an interfacial 'slip-layer' the metal and channel wall. Thereafter, an applied electric field (~10-20 V/m) actuates the liquid metal by establishing a gradient of surface tension; this effect is known as continuous electrowetting (CEW). Although CEW has been utilized before with mercury, which is toxic, the adhesive nature of the Ga oxide complicates CEW behavior. This work utilizes optical microscopy, cyclic voltammetry, and electrochemical impedance spectroscopy (EIS) to characterize electro-hydrodynamics of the system under a variety of conditions. Specifically, we elucidate the influence of electrolyte (i.e., composition, pH, and viscosity) on the metal-electrolyte interface. In addition, we compare electro-hydrodynamic behavior of EGaIn with and without the presence of an oxide 'skin.' Finally, this work explores novel microfabrication strategies to design interfaces that prevent oxide adhesion.

A Model for Electrokinetic Flow with Deformable Interfaces

Michael Booty[1], Rui Cao[1], Manman Ma[2] and Michael Siegel[1], (1)Mathematical Sciences, New Jersey Institute of Technology, Newark, NJ, (2)Mathematical Sciences, Tongji University, Shanghai, China

Abstract

A model is introduced to describe the deformation of a fluid drop in the two-phase flow of immiscible, ionic fluid electrolytes in the presence of an impulsively applied DC electric field.

The starting point is the Poisson-Nernst-Planck equations in the zero Reynolds number or Stokes flow regime for symmetric binary electrolytes. These are analyzed by introducing matched asymptotic expansions in the limit when the thickness of the induced charge Debye layers that form adjacent to the interface is much less than the initial size of the undeformed drop. This leads to a hybrid or multiscale model that consists of boundary integral equations for the electrostatic potential and the interface fluid velocity together with relations that contain the coupling between the electrostatic and fluid fields within the thin Debye layers.

The drop interface is assumed to be sharp, i.e. of zero thickness relative to the scale of the Debye layers, and the induced charge resides in the Debye layers with no surface charge on the sharp drop interface. A motivation for the study is to understand the influence of bulk electrokinetic effects on interfaces that are deformable and can evolve in time, in the simple context of absence of surface charge, while accurately resolving both Debye layer structure and interface shape and dynamics.

Dimensionless groups that appear in the model include the ratio of material properties in the interior and exterior electrolyte phases for the electrical permittivity, available molar ion concentration, the fluid viscosity, and the ion diffusivities. The typical speed at which ions diffuse across the Debye layers is the velocity scale, which appears in a dimensionless ratio with the interfacial capillary velocity. The main control parameter is the applied voltage, which appears in dimensionless ratio with the thermal voltage and in a second, independent ratio of the scaled electrostatic stress to viscous stress.

The numerical implementation of the model and results of the numerical simulations will be presented, together with comparison to a small-deformation analysis and discussion of the dependence of the results on system parameters.

This work was supported in part by NSF grant DMS-1412789.

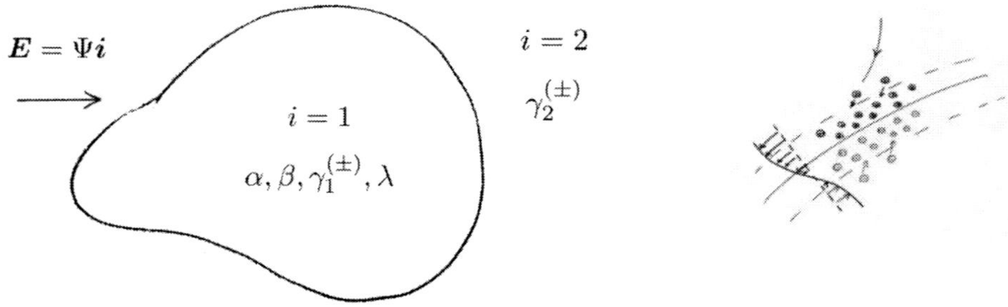

$$0 < \epsilon = \frac{\lambda_D}{a} \ll 1, \quad \Psi = \frac{E_\infty a e}{kT}, \quad \Delta = \frac{\sigma}{\mu_2 U^*}, \quad \nu = \frac{\varepsilon_2 \Phi^{*2}}{\mu_2 D}$$

$$U^* = \frac{D}{\lambda_D} = U_c \ \textit{is the charge-up speed}, \quad \Phi^* = \frac{kT}{e}$$

$$\Rightarrow \ Pe = \frac{U^* a}{D} = \epsilon^{-1} \gg 1, \quad Ma^{-1} = \frac{\varepsilon_2 \Phi^{*2}}{\mu_2 U^* a} = \frac{U_{HS}}{U_c} = \nu \epsilon$$

Poisson-Nernst-Planck equations with Stokes flow and a sharp interface

$$\textit{ion conservation}\quad (\partial_t + \boldsymbol{u}\cdot\nabla)c^{(\pm)} - \epsilon\gamma_i^{(\pm)}\nabla\cdot(\nabla c^{(\pm)} \pm c^{(\pm)}\nabla\phi) = 0$$

$$\textit{molecular ion flux}\quad \boldsymbol{j}^{(\pm)} = -\epsilon\gamma_i^{(\pm)}(\nabla c^{(\pm)} \pm c^{(\pm)}\nabla\phi)$$

$$\textit{interface conditions (i):}\quad \boldsymbol{j}^{(\pm)}\cdot\boldsymbol{n} = 0, \quad [\alpha_i\nabla\phi\cdot\boldsymbol{n}] = 0$$

$$\textit{Poisson}\quad \epsilon^2\nabla^2\phi = -\frac{\beta_i}{\alpha_i}q, \quad q = \frac{c^{(+)} - c^{(-)}}{2}$$

$$\textit{Stokes flow}\quad \nabla\cdot\boldsymbol{u} = 0, \quad -\nabla p + \lambda_i\nabla^2\boldsymbol{u} + \nu\epsilon\alpha_i\nabla^2\phi\nabla\phi = 0$$

$$\textit{interface conditions (ii):}\quad [(\boldsymbol{T}_H + \nu\epsilon\boldsymbol{T}_M)\cdot\boldsymbol{n}] = \Delta(\kappa_1+\kappa_2)\boldsymbol{n}$$
$$[\boldsymbol{u}] = 0, \quad (\frac{d\boldsymbol{x}_s}{dt} - \boldsymbol{u})\cdot\boldsymbol{n} = 0$$

$$\textit{far-field:}\ \phi \sim -\Psi x, \quad c^{(\pm)} \to 1, \quad \boldsymbol{u} \to 0$$

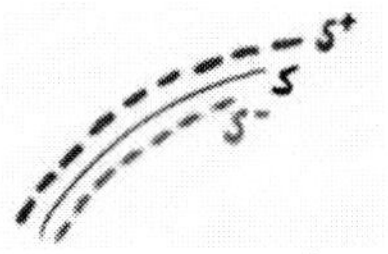

$$\textit{Outside the layers}\quad (\partial_t + \boldsymbol{u}\cdot\nabla)c = 0 \quad \textit{where}\quad c = \frac{c^{(+)} + c^{(-)}}{2}$$
$$\nabla^2\phi = 0, \quad \nabla\cdot\boldsymbol{u} = 0, \quad -\nabla p + \lambda_i\nabla^2\boldsymbol{u} = 0$$

$$\textit{Debye layer conditions:}\quad \zeta_+ - \zeta_- = -[\phi]_{S_-}^{S_+} \quad \sqrt{c}|_{S_+}\sinh\frac{\zeta_+}{2} + \sqrt{\alpha\beta c}|_{S_-}\sinh\frac{\zeta_-}{2} = 0$$

$$\begin{aligned} \textit{On } S_+\textit{:}\quad c\partial_n\phi &= -(\partial_t + \boldsymbol{v}_s\cdot\nabla_s - \partial_n v_p|_s)\sqrt{c}\left(\frac{e^{\zeta_+/2}-1}{\gamma_2^{(-)}} - \frac{e^{-\zeta_-/2}-1}{\gamma_2^{(+)}}\right) \\ \textit{if }\ \partial_n v_p|_s > 0\ \ \partial_n c &= (\partial_t + \boldsymbol{v}_s\cdot\nabla_s - \partial_n v_p|_s)\sqrt{c}\left(\frac{e^{\zeta_+/2}-1}{\gamma_2^{(-)}} + \frac{e^{-\zeta_-/2}-1}{\gamma_2^{(+)}}\right) \end{aligned}$$
$$\textit{where}\quad \partial_n v_p|_s = -(\nabla_s\cdot\boldsymbol{u}_s + (\kappa_1+\kappa_2)\boldsymbol{u}_s\cdot\boldsymbol{n})$$

On S_-: similar. Stress-balance BC depends on $([p], \boldsymbol{u}_s, c|_{S_\pm}, \zeta_\pm)$

An integral equation for $\boldsymbol{u}$

$$\boldsymbol{u}(\boldsymbol{x}_0) = -\frac{1}{4\pi(1+\lambda)}\int_S \boldsymbol{G}(\boldsymbol{x},\boldsymbol{x}_0)\cdot[\hat{\boldsymbol{T}}\cdot\boldsymbol{n}(\boldsymbol{x})]dS_{\boldsymbol{x}}$$
$$+\frac{(1-\lambda)}{4\pi(1+\lambda)}\int_S^{PV} \boldsymbol{u}(\boldsymbol{x})\cdot\boldsymbol{T}(\boldsymbol{x},\boldsymbol{x}_0)\cdot\boldsymbol{n}(\boldsymbol{x})dS_{\boldsymbol{x}}$$
$$-\frac{\nu}{4\pi(1+\lambda)}\int_S \boldsymbol{n}(\boldsymbol{x})\cdot\boldsymbol{\nabla}_{\boldsymbol{x}}\boldsymbol{G}(\boldsymbol{x},\boldsymbol{x}_0)\,\Theta(\boldsymbol{x})\,dS_{\boldsymbol{x}}$$

$$\textit{where}\quad \Theta = 4\left(\sqrt{c}|_{s_+}\sinh^2\frac{\zeta_+}{4} + \sqrt{\alpha\beta c}|_{s_-}\sinh^2\frac{\zeta_-}{4}\right)$$

$$\textit{and}\quad [\hat{\boldsymbol{T}}\cdot\boldsymbol{n}] = (\kappa_1+\kappa_2)\,(\Delta-\nu\Theta)\,\boldsymbol{n} + \nu\nabla_s\Theta$$

An integral equation for ϕ

Time-dependent small deformation analysis

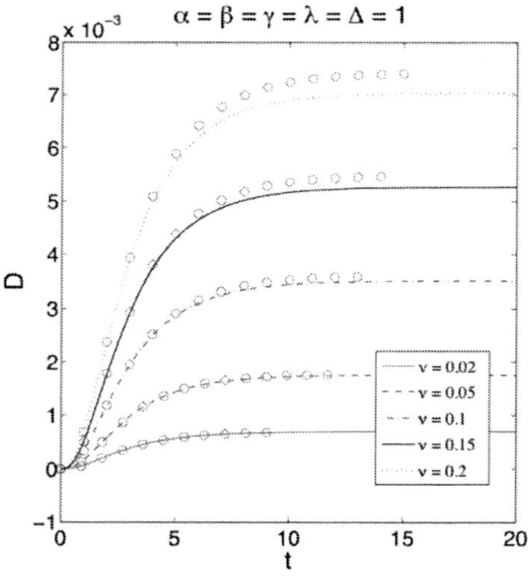

$$D \simeq \frac{9(\lambda+4)\sqrt{\alpha\beta}\,\nu}{320(\lambda+1)(1+\sqrt{\alpha\beta})\Delta}\Psi^2$$

Validation of a Legendre pseudospectral method for the 'full' governing equations with the small-ε study of Schnitzer et al. Electrophoresis of bubbles, JFM 2014

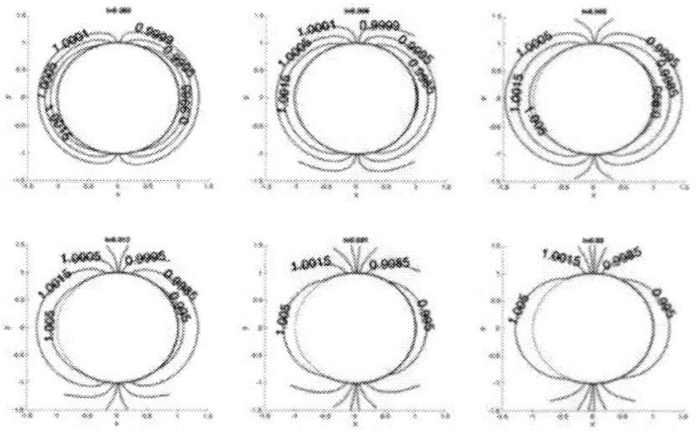

Development of the total ion concentration, or c-field, with time

492400 Monodispersed Droplet Generation Using AC Electric Field

Abstract:

We report a new microfluidic technology that uses an AC electric field to generate mono-dispersed water-in-oil droplets that range from 10 to 100 Î¼m in diameter, without a conventional flow focusing design. The droplet size of this technolocy can be easily tuned by adjusting the voltage/frequency and is hence much easier to use then flow-focusing designs, which adjust water/oil flow rate ratios to vary the droplet size. As such, it is ideal for multiplex digital PCR and high-throughput emulsification applications that require frequent size tuning for a large dynamic range.

The droplet generation unit is a micropipette within a tubing with water being pumped through the micropipette and a co-flowing oil in the annular region to convect the water droplets away downstream. By using laser-assisted pulling, the tip of the micropipette is pulled into a conic shape with a tip diameter of about 10 Î¼m[1]. This conic geometry focuses the electric field so that the Maxwell pressure is sufficient to overcome capillary pressure even at small scales. The oil phase contains surfactants that stabilizes the emulsion but also allows droplets to pinch off at the tip of the micropipette.

Unlike flow focusing, our field focusing design does not produce a jet that breaks up by the Rayleigh instability or by a convective version of the instability that pinches off the tip of a long jet. We also avoid large droplets at the tip, like those pinched off by high oil shear rates at T-junctions, whose critical viscous drag for pinch-off is only reached for droplets much larger than the capillary tip. Instead, we tune the frequency of the AC field such that a Maxwell pressure is imparted over a large area of the first droplet, without producing a cone or jet[2]. This large pressure accelerates the drop forward at a rate that cannot be matched by the water feed flow rate, much like inertial acceleration but at negligible feed Reynolds numbers. As in inertial acceleration of a finite-volume drop, a neck forms at the capillary tip that pinches off single droplets at a very high rate beyond 100Hz. Phase diagrams are constructed in the parameter space of dimensionless voltage (Bond number), frequency, flow rate (Weber number), viscous drag and surface tension (capillary number). Correlations for the droplet size and generation frequencies in terms of these dimensionless parameters will also be reported to collapse the measured data.

[1] Pan, Zehao, et al. "Universal Scaling of Robust Thermal Hot Spot and Ionic Current Enhancement by Focused Ohmic Heating in a Conic Nanopore." Physical Review Letters 117.13 (2016): 134301.

[2] Chetwani, N., Maheshwari, S. and Chang, H.-C."Universal cone angle of ac electrosprays due to net charge entrainment", Phys Rev Lett, 101, 204501(2008).

498927 Crater Formation on Electrodes during Charge Transfer with Aqueous Droplets or Solid Particles

Abstract:

We report that metallic electrodes are physically pitted during charge transfer events with water droplets or other conductive objects moving in strong electric fields (>1 kV/cm). Post situ microscopic inspection of the electrode shows that each individual charge transfer event yields a crater approximately 1 to 3 microns wide, often with features similar to splash coronae. We interpret the crater formation in terms of localized melting of the electrode via resistive heating concurrent with dielectric breakdown and charge transfer through the surrounding insulating fluid. A scaling analysis indicates that the crater diameter scales as the inverse cube root of the melting point temperature T_m of the metal, in accord with crater measurements on several metals (660âµ'C $< T_m <$ 3414âµ'C). The process of crater formation provides a possible explanation for the longstanding difficulty in quantitatively corroborating Maxwell's prediction for the amount of charge acquired by spheres contacting a planar electrode.

499536 The Electric Field in Water between Parallel Electrodes: A Sinusoidal Applied Potential Can Yield a Non-Zero, Long-Range Steady Field

Abstract:

We use the standard electrokinetic model to numerically investigate the electric field in aqueous solutions between parallel electrodes under AC polarization. In contrast to prior work, we invoke no simplifying assumptions regarding the applied voltage, frequency, or mismatch in ionic mobilities. We find that the nonlinear electromigration terms significantly contribute to the overall shape of the electric potential vs. time, which at sufficiently high applied potentials develops multi-modal peaks. More surprisingly, we find that electrolytes with non-equal mobilities yield an electric field with non-zero time average at large distances from the electrodes. Our calculations indicate this long-range electric field suffices to levitate colloidal particles many microns away from the electrode against the gravitational field, in accord with experimental observations of such behavior (Woehl et al., PRX, 2015). Moreover, the results suggest that objects in water will rarely experience a sinusoidal applied electric field despite application of a perfectly sinusoidal applied potential, calling into question many theoretical calculations predicated on this assumption.

498911 Moving Past Simple Shapes: Engineered Active Particle Spinners and Motors Powered By AC Electric Fields

Abstract:

Active particles have gained tremendous attention in recent years due to their ability to harvest energy and autonomously power their movement, organization and reconfiguration. Ranging from molecular motors on the nanoscale to living organisms on the macroscale, active particles exist outside of equilibrium to convert energy into motion. The underlying mechanism driving their movement is in their design, as they asymmetrically draw and expel energy, thus creating local gradients of force for propulsion. While nature provides abundant examples of active particles that display highly advanced locomotive behaviors, progress in the lab has been limited by constraints in particle fabrication, such that most active particles developed insofar are simple Janus spheres. This bottleneck has restricted the potential for realizing new types of active particles with advanced properties for executing complex and intricate tasks. In this talk, we will show that the design and movement of active particles can be advanced through leveraging concepts in monolithic fabrication. We will show that a rich variety of particle designs with complex geometries (i.e., well-defined, non-spherical shapes) and polarizabilities can be achieved by top-down approaches (e.g., multi-layered lithography and metal evaporation). This results in the formation of millions of monodisperse, polymeric particles comprising a rich variety of shapes and metallic patches that give rise to a number of interesting and programmable field-particle and particle-particle interactions. Inspired by natural systems, we will show how this new class of â€œengineered active particlesâ€ can be designed in a variety of ways to swim, spin and freeze/melt on demand and in a tunable fashion. We will focus on the influence of shape and polarizability on the motile responses of the particles in an applied AC electric field. We will also investigate the collective behaviors of some of these particles, and we will provide descriptive physical explanations for their locomotion based on the interplay and coupling of electrohydrodynamic flows and induced charged electro-osmotic flows. Finally, building off of this understanding, we will show how the responses of the engineered particles can be enhanced or more finely controlled by leveraging the metallic patches along specific regions of the particles. Together, this work may provide a foundation for the next generation of active particles with engineered properties to execute a wide range of tasks, including remotely powered vehicles for targeted drug delivery, reconfiguring particle assemblies for microsurgical devices and swarms of particles for the next generation of self-healing materials.

490539 Entropic Trap-Based Tunable Short-Pass Filter to Recover Long DNA for Genomic Applications

Abstract:

Long DNA molecules are of extreme importance for genome technologies such as long-read sequencing, whole-genome mapping and cell engineering. In many situations, the long DNA sample is contaminated by a significant amount of small DNA that results from random shearing or biochemical processing. These small DNA fragments contribute noise to the analysis; for example, genome mapping in nanochannels requires DNA greater than 150 kbp and small DNA can create spurious signals when they overlap with the larger molecules. Gel electrophoresis is the standard method to purify long DNA molecules, but the method is time consuming, due to the slow mobility of long DNA in a gel, and requires a high initial DNA concentration for purification. Here, we demonstrate that entropic trap-based filters, fabricated by reactive ion etching in a fused silica substrate, can selectively trap long DNA at a nanoslit-microchannel interface while passing the smaller DNA. The trapping ability of this short-pass filter can be tuned by optimizing the strength of the electric field at the interface and the relative height of nanoslit to the radius of gyration of the DNA. Our approach also involves designing an automated, continuous separation system that decouples the injection and filtration steps by using air pressure and electric potential, respectively. This tunable short-pass filter system provides an efficient alternative for gel electrophoresis for sample preparation of rare long DNA molecules, which can be integrated to lab-on-chip devices for next-generation genome technologies.

498792 Dielectrophoretic Quantification of Mixed Blood Populations for Detection of Autologous Blood Transfusions

Abstract:

The desire to increase performance in professional athletes has driven some to turn to artificial methods to achieve this end. Autologous blood transfusions – the collection and subsequent reinfusion of a person's own blood – currently lack a direct test for detection, and has for that reason become a popular "banned method" in performance enhancement. The Athlete's Biological Passport is meant to monitor an athlete's long-term trends in numerous biomarkers, and use abnormal variations in the markers to potentially identify blood-doping. It does provide an indirect means of identifying autologous blood doping, but there still lacks a direct detection method.

Using a combination of membrane protein cross-linking and a microfluidic device designed for isolated dielectrophoretic spectroscopy, we demonstrate the ability to detect the presence of 5% reinfused red blood cells (RBCs) in fresh blood.

Red blood cells aging in storage undergo biochemical changes that normal RBCs in circulation do not. This includes changes in cell membrane composition and cell morphology. When stored ex vivo, RBCs begin to lose the number of membrane-bound proteins. This physiological change offers less surface proteins available for cross-linking, resulting in a different dielectrophoretic spectra for stored cells compared to that of fresh RBCs. We use differences in these spectra to discern storage-related changes to blood cell morphology and composition, as a means to directly detect autologous blood doping in athletes.

Further, we show that accurate DEP quantification of these RBC subpopulations requires analysis chambers with confined micron-scale heights. This is because both gravity and the non-uniformity in the electric field gradient produce DEP separation ratios that do not reflect the actual cell mixture percentages. To alleviate this challenge, we present a novel confined DEP design that increases the separation and quantification accuracy of the blood sample for detection of autologous blood doping in endurance athletes.

498699 Non-Optical Biomolecular Detection in Human Serum Using Interfacial Electrokinetic Transduction

Abstract:

Development of rapid, sensitive and portable detection systems are important for effective detection of diseases in developing countries, biowarfare/anti-terrorism applications, environmental monitoring, and for basic biological research. Biosensors are analytical devices that detects an analyze of interest by combining biological sensing elements with physiochemical transducers. While current biosensing methods are capable of sensitively detecting a variety of target analytes in a laboratory setting, there are inherent difficulties in developing low-cost portable biosensors for point-of-care diagnostics using traditional optical, mass, or electroanalytical-based signal transducers. It is therefore important to develop new biosensing transducer elements for recognizing binding events at low cost and in portable environments. This approach utilizes a novel electrical method that combines a laminar liquid-liquid interact with alternating current (ac) electrokinetics. A bioreaction occurs at the co-flowing liquid interface with an ac electric field is simultaneously applied, displacing the interface across the flow channel. The magnitude and direction of displacement is directly related to both the applied frequency and interfacial electrical properties. Biosensor detection is based on interfacial electrokinetic transduction; binding changes the interfacial electrical properties, which is electrically traduced and detected by monitoring the frequency response of the liquid interface. In this manner, the liquid interface serves as both the substrate for specific binding, and transducer for biomolecular recognition. Furthermore, by implementing electrical impedance spectroscopy (EIS), the interfacial position can be monitored non-optically. Utilizing EIS, the biosensor becomes label-free and non-optical; the magnitude of impedance, which can be monitored on a computer screen, becomes the readable signal. Here, a non-optical, label-free biosensor, which can detect bimolecular reactions in human serum is created by pairing interfacial electrokinetic transduction and electrical impedance spectroscopy.

486888 Pulsed-Field Electrophoresis for Microfluidic Devices

Abstract:

Microfluidic device provides many advantages including portable device, low sample consumption, fast separation and short analysis time. In electrophoresis, the motion of particles with different charge and size will lead to separation of different analytes owing to the different electrophoretic mobility under the electric field. One advantage of pulsed field is that it can apply appropriate periodic field to enhance separation. Here, we employed the timing pulsed electric field on microchip, to enhance the separation of free solution electrophoresis. A microchip made of zeonor with a cross channel was used to demonstrate separation of two analytes, which were negative and neutral dyes. The effect of injection, and separation steps were programmed by using a high voltage sequencer. The separation was studied with laser induced fluorescence. CCD camera was used to monitor and record the images for analysis. Fast separation was observed. Furthermore, the pulsed field has higher signal intensity and less dispersion of sample, than the continuous field does. The results indicated that pulsed field could reach the high speed separation with relatively low voltage, reduce dispersion and improve the signal to noise ratio, and resolution, compared with continuous field.

486887 Low-Loss on-Chip Sample Pretreatment: Depletion Isotachophoretic Isolation of DNAs from Inhibitor-Rich Samples By an Ionic Transistor

Abstract:

Low-copy number DNA detection from clinical samples is plagued by analyte loss during the various personnel-intensive sample pretreatment steps like fractionation, centrifugation and dialysis to isolate the target molecules and to remove the inhibitors or interfering agents for PCR amplification or the particular biochemical assay. In order to eliminate the undesired sample handling and simplify the sample pretreatment steps, we report an on-chip sample pretreatment device to isotachophoretically isolate DNA molecules from arbitrary inhibitor-rich samples. DNA molecules with a particular length can be concentrated and purified, without confining barriers, at a specific on-chip location for easy transfer to on-chip PCR amplification units and downstream assays.

Our sample pretreatment device realizes depletion isotachophoresis by a microfluidic ionic transistor. The microchannel for DNA isolation is gated by an ion-selective nanoporous membrane. Because of the charge selectivity of the ion-selective membrane, ion concentration depletion occurs in the microchannel upon the application of a normal gating electric field across the membrane[1]. By configuring the gating and draining voltages applied on the ionic transistor, we can achieve a gating voltage-sensitive but stationary depletion front length in the microchannel. Such a stationary depletion front offers a stable electric field jump to isotachophoretically isolate charged molecules, like DNAs, at a particular location selected by the gating voltage.

By continuously loading samples into the ionic transistor, we can isolate negatively charged DNA molecules at the stationary depletion front from samples with an arbitrary composition and volume. Inhibitors with positive charges or with low electrophoretic mobility can be removed by the electric field and the loading flow simultaneously, without the need of centrifugation and filtration. By real-time fluorescence imaging, we show that DNA molecules with different lengths, ranging from 30 bases to 400 bases, are isolated into separated bands with a near 90% isolation yield. The DNA concentration in each isolated band is also enhanced by two orders of magnitude. We further prove that the pretreatment device removes PCR inhibitors, including Calcium ions and Hemoglobin, by successfully performing on-chip PCR of the isolated DNAs. Integrating the pretreatment device with the on-chip PCR amplification and an ion current nanomembrane sensor developed in our lab for DNA detection[2], we expect to achieve multiplex nucleic acid sensing with fM sensitivity directly from serum and urine samples.

[1] Chang et al, Annu Rev Fluid Mech, 44, 401-426 (2012)

[2] Slouka et al, Annu Rev Anal Chem, 7, 317 (2014)

497323 Plasmonic ELISA Biosensor with Tunable Sensitivity and Selectivity

Abstract:

In todayâ€™s connected world, rapid, cost-effective, transformational, operator-friendly biosensors are the need of the hour to prevent sudden outbreaks, disease spread or for preventive intervention. This has lead to extensive research in biosensors, firmly divided into two important subclasses, namely, the point-of-screening devices and point-of-care devices. Here we describe a tunable device that uses current well-established device architecture like ELISA, plate readers and translates it into a faster, cheaper, more selective and more sensitive biosensor. The tunability of our Biosensor allows it to be used as both, point-of-screening or as a point-of-care device. Our biosensor architecture is composed of two 500 Î¼m X 100 Î¼m thick channels separated by a gold coated polycarbonate membrane. Interdigitated electrodes with a width of 10 Î¼m and separated by a pitch of 10 Î¼m enclose the channels. The porous gold membrane allows for multiple activities. Namely, the binding of the antibody to the gold membrane, the passage of the solution and the unique repetitive structure of the membrane allows for plasmonic enhancement of any optical signal. A DNA with a hairpin structure is used for fluorescent characterization of our gold coated membrane-based biosensor to quantify the plasmonic enhancement. The previous literature states that proximity to metal surface leads to quenching while plasmonic enhancement decreases with increasing distance. The distance between the fluorescent dye and the membrane surface is altered using a molecular separator (ssDNA and thiol linkers). This allows us to optimize the plasmonic enhancement curve with distance. The enhancement results are further verified using finite difference time domain simulation using commercial software. Glancing angle deposition (GLAD) of metals on a regular well-ordered nanometer-sized architecture lead to nano-ordered hierarchical structures. These hierarchical structures can result in local surface plasmon resonance (LSPR) which can cause tremendous enhancement of the electric field. Using our membrane as the nanometer-sized architecture, we deposited gold at multiple different GLAD angles from 0° to 75°. Each GLAD angle deposition leads to unique nano-ordered hierarchical structures that were analyzed using AFM and SEM. The fluorescent enhancement from each structure was quantified and the results matched to theoretical simulations. We performed the well-established sandwich ELISA to compare the performance of our biosensor against those found in the literature. We concluded that our biosensor was more selective and more sensitive than those in literature. Finally, the AC electric field between the interdigitated electrodes and the gold coated membrane will be modulated to observe the effect of the AC frequency on the binding chemistry and the orientation of the antibody linked to the gold membrane. As has been illustrated elsewhere, the electric field can be used to orient the antibody leading to significant increase in the sensitivity and selectivity of the antibody. By changing the orientation of the antibody in the fluidic field with respect to the pores of the membrane, the selectivity of the antibody can be altered. Interestingly, the orientation of the antibody also allows to either diminish the fluorescence from the molecule (quenching from the gold coated membrane) or increase the fluorescence enhancement (minimal quenching and maximal plasmonic). This allows us to modulate the sensitivity and selectivity of the biosensor in real-time and hence the tunability of our Biosensor.

497628 Modeling and Validation of the Effect of Electric Field on Drug Delivery into the Tumor Cell

Abstract:

A tumor microenvironment has complex multiphysics and multiscale characteristics. To deal with this we propose the use of agent-based modeling (ABM). Specifically, ABM is employed to predict the tumor cell behavior in contact with a chemical perturbagen or drug of interest; this is essential for personalizing cancer treatment. There is the fact that the tumor microenvironment is very heterogeneous due to angiogenesis, creating significant barriers for effective drug delivery. Application of an electric field has the potential to overcome some of these barriers. Equation-based computational models have been developed to study electrokinetics phenomena. However, to our knowledge, in the open literature, an ABM has not been used to study the effect of applied electric field on a complex chemotherapeutic delivery system. In this paper we discuss the development of an ABM model that would lead to a significantly more realistic prediction of the optimum electric field and chemotherapeutics that would result in the maximum fraction of cell killed. We also used COMSOL Multiphysics simulation to solve equation-based computational model that simultaneously considers all three electrokinetics phenomena (electrophoresis, electroporation, and electroosmosis). The simulation is based on the coupling of Smoluchowski, Nernst-Planck for three specific species, and Navier-stokes equations. Finally, we have simulated the microfluidic device for validating the deterministic model.

494636 A Shear-Enhanced CNT-DEP Nanosensor Platform for Single Cell Protein Assay

Abstract:

Breast cancer is the main cause of cancer deaths among women, with approximately 1.4 million new cases and nearly 0.5 million deaths worldwide per year. As the growth of tumor varies from patients to patients, its mortality rate can be reduced if the hormone receptor status of ER, PR and HER2 can be accurately determined from cells within tumors. Since tumor cells are heterogeneous, it is desirable to analyze individual cell rather than an average bulk measurement in order to extract accurate information. However, current high-throughput and quantitative technology Reverse Phase Protein Array (RPPA) still require at least 10 cells equivalents of sampleÂ [1]. Hence, there is a need to develop a detection platform for single-cell proteomics that will enable precise treatment plan based on individualâ€™s specific information within tumor cells and significantly reduces the side effects of therapeutics.

We report a new nanoscale protein sensor platform that uses irreversible electrokinetic phenomena to achieve single cellular sensitivity and specificity for HER-2 protein assay. The sensitivity is achieved with irreversible rare events, driven by a DEP (dielectrophoresis) driven (antibody-antigen-antibody) Ab-Ag-Ab ELISA complex association, which can nevertheless be detected with a sensitive CNT electron tunneling conductance sensor design. The selectivity is achieved by working at a critical shear rate that is between the dissociation shear rate of the target and the non-target. Specific components of this new protein sensing platform include: 1. Association of protein targets with Ab probes functionalized onto an gold electrode pair; 2.Transportation and trapping of the second linking Ab probe functionalized CNTs with long-range DC electrophoresis and short-range AC DEP to the electrodes (lock); 3.Selective removal of assembled CNTs with non-targets by an optimized nanoshear protocol; 4.Conductance quantification of remaining bridging CNTs with target linkers (switch). Steps 1 and 2 involve rapidly driven dynamic events to prevent the captured targets from dissociating from the probe to reach the thermodynamic coverage. Optimized shear at the nm-high hydrodynamic slip length in step 3 irreversibly removes CNTs with non-targets due to the large Stokes drag of their high aspect-ratio cylindrical geometry. With only target-linked CNTs remaining, the digital conductance by electron-tunneling across the Ab-Ag-Ab complex allows us to reach a detection limit of 10,000 molecules (10 fM) for spiked HER2 samples. Irreversible capture and shearing also allow us to tune the dynamic range up to 100 billion (100 pM or 5 decades) by increasing the CNT number. The spiked sample results also demonstrated high selectivity towards target HER2 proteins against non-target HER2 isoform of a similar K_D. We will also carefully conduct experiments with real tumor cell lysate, and the number of proteins detected will be correlated with respect to tumor cell numbers.

References:

1. Neeley ES, Baggerly KA, Kornblau SM, Surface Adjustment of Reverse Phase Protein Arrays using Positive Control Spots, *Cancer Inform.*, 2012; 11; 77-86.

503991 Flow Regulated Anodic Growth of TiO_2 Nanotubes in Microfluidics

Abstract:

Electrochemical anodization of titanium (Ti) in a static, bulk condition has been used widely to fabricate self-organized TiO_2 nanotube arrays. Such bulk approaches, however, require extended anodization time to obtain long TiO_2 nanotubes and produce only vertically-aligned nanotubes. To date, it remains challenging to develop effective strategies to grow long TiO_2 nanotubes in a short period of time and control the nanotube orientation. Here, we show that the anodic growth of TiO_2 nanotubes is significantly enhanced (~16-20 times faster) under flow conditions in microfluidics. Flow not only controls the diameter, length, and crystal orientations of TiO_2 nanotubes but also regulates the spatial distribution of nanotubes inside microfluidic devices. Strikingly, when a Ti thin-film is deposited on silicon substrates and anodized in microfluidics, both vertically- and horizontally-aligned (relative to the bottom substrate) TiO_2 nanotubes can be produced. Our results demonstrate previously unidentified roles of flow in the regulation of growth of TiO_2 nanotubes and provide powerful approaches to effectively grow long oriented TiO_2 nanotubes and construct hierarchical TiO_2 nanotube arrays on silicon-based materials.

503702 Dielectrophoretic Separation of Large Microscale Particles (dp>5 um) By Exploiting Charge Differences

Abstract:

Electrokinetics is the family of phenomena that depend on the electrical double layer. Electrokinetic techniques are one of the main pillars of microfluidics, due to their ease in application. Electric-field driven technique such as electroosmosis, electrophoresis and dielectrophoresis have been successfully used for the analysis, sorting and separation of a wide array of bioparticles, in applications than range from environmental assessments to biomedical and clinical analysis. In this current project, efforts have been made to analyze the equilibrium between the electrophoretic, electroosmotic and dielectrophoretic forces and how this equilibrium is affected by particle size. As the motion of larger particle under electric fields has yet to be fully characterized, the primary focus has been particles with diameters ranging from 5-10 microns. In theory, these "larger" particles should be easier to "trap" in our insulator-based dielectrophoresis (iDEP) systems when compared to smaller particles, since DEP force depends on particle volume, however, this is not the case. Throughout this study an iDEP microchannel, with an array of cylindrical insulating structures and direct current electric fields have been employed. Results from current and preliminary experiments have showed that larger carboxylated polystyrene particles (diameter > 5 mm) require much higher voltages than expected and also have shown to move very fast through such iDEP systems. Employing suspending media with conductivity of 15-20 mS/cm and pH of 6-7, under applied fields between 400 V/cm and 1500 V/cm, the 5-mm, 7-mm and two types of 10-mm polystyrene particles were observed to become immobilized due to negative electrophoretic trapping. In addition, it our experiments have reveled that the amount of surface charge of the particles may also have an impact of the ability to trap these larger particles and will be further explored as the project moves forward. The results of this project will hopefully be able to help explain why these larger particles do not behave according to theory. The recent finding in reference to amount of surface charge also has the potential to become another method by which biological particles can be separated through the use of iDEP.

503704 Dielectrophoretic Assessment of Sub-Micron Particles By Exploiting Charge Differences

Abstract:

Dielectrophoresis (DEP) has an immense potential for particle sorting and separation in microscale systems. Specifically, insulator-based dielectrophoresis (iDEP) devices are both inexpensive and effective for the analysis of biological particles or cells. In previous work, it has been seen that submicron carboxylated particles (dp < 500 nm) experience a positive DEP force when exposed to a non-uniform electric field, while larger particles (dp>1 mm) experience a negative DEP force [1]. In this research, the effect of particle surface charge magnitude on the DEP force is investigated. Two different particle surface functionalizations are also analyzed: carboxylated and aminated (negatively and positively charged, respectively). Being able to separate sub-micron particles by exploiting charge differences offers great potential in bioanalysis.

1. Mario A. Saucedo-Espinosa Mallory M. Rauch Alexandra LaLonde Blanca H. Lapizco-Encinas, "Polarization behavior of polystyrene particles under direct current and low-frequency (<1 kHz) electric fields in dielectrophoretic systems," *Electrophoresis* 2016, 37, 635–644

496033 Research of DNA Separation By Post Array Under Intermittent Electric Field

Abstract:

We applied an intermittent electric field to replace the continuous one to separate λ-(48.5kbp) and T4(165.6kbp) DNA in a fused silica microchannel with a hexagonal post array of 1 micron diameter and 3 micron pitch. In the post array, the mobility of DNA decreases with molecular weight since larger DNA has higher probability to hit and hook on the post. However, the occurrence of channeling phenomenon restricts not only the operating condition but also the separation efficiency of a fixed length post array. By using intermittent electric field, the electric field(E) is repeatedly turned on for a time interval t_{on} and turned off for a time interval t_{off}. With an intermittent electric field, the channeling phenomenon was suppressed by the relaxation of DNA during the "off" period. We also correlated t_{on} with the trapping time of DNA. The results indicated that t_{on} set in between the trapping time of T4 and λ-DNA has the best resolution of separation for a fixed separation time while t_{on} set close to the trapping time of λ-DNA yields the best resolution of separation for a fixed channel length. Compared to continuous electric field, an intermittent one enables the separation to be conducted under higher Pe. With properly tuned t_{on} and E, the resolving power of a channel with fixed length increases only at the expense of time. For a given post array, intermittent electric field offers much more flexibility in choosing operation conditions for optimized separation.

496257 Insight into Coal Structure Based on Benzene Carboxylic Acids from the Coal Via Oxidation

Abstract:

Insight into coal structure based on benzene carboxylic acids from the coal via oxidation

Fan Yang[1], Yucui Hou[2], Muge Niu[1], Shuhang Ren[1], Weize Wu[1,*]

[1]State Key Laboratory of Chemical Resource Engineering, Beijing University of Chemical Technology, Beijing 100029, China

[2]Department of Chemistry, Taiyuan Normal University, Taiyuan 030031, China

Coal structures, the basis of its applications, have been widely studied. Benzene carboxylic acids (BCAs, 12 types) can be generated from different aromatic clusters of coal via oxidation, and it has been found the 12 BCAs have different yield distributions for different coals. The results suggest that there is a certain relationship between BCAs and coal structure that provides a favorable basis for us to study the coal structures. In this work, based on the BCAs distributions and ^{13}C-NMR analyses, we investigated the structure characters of coal with different ranks. The results indicate that with the increase of coal rank, the yield of BCAs increase, and the structures of coal becomes more and more difficult to be degraded. There are great differences in the yield distributions of BCAs with the increase of coal rank, and more and more BCAs with a small number of carboxyl are obtained in high rank coal. Single-ring aromatic clusters and double-ring aromatic clusters are mainly aromatic structures in low rank coal. With an increase in the process of coalification, the aromatic clusters increase in size and the degree of condensed of aromatic rings also increases. When the carbon content of coal was approximate 87 %, the structure of coal had a mutation. At last, we proposed a structural model of Huolinhe lignite based on the yield distribution of BCAs. The work provides a new way to study the coal structures and construct the structural model of coal.

The project is financially supported by the National Natural Science Foundation of China (No. 21676019 and 21076138).

Keywords: Coal structure; Benzene carboxylic acid; Oxidation; ^{13}C-NMR

---

* Corresponding author; E-mail: wzwu@mail.buct.edu.cn; Tel./Fax: +86 10 64427603.

498379 Multiphysics Modeling of Microfluidic Device to Investigate the Effect of Electric Field on Drug Delivery into the Tumor Cell

Abstract:

Studying the uptake rate of the macromolecules under the effect of electrokinetics phenomena (electrophoresis, electroporation, and electroosmosis) has an application in personalizing electrochemotherapy treatment. The uptake rate of chemotherapeutics can be a marker for single cell characterization. In this study, the COMSOL Multiphysics simulation to solve equation-based computational model on coupled electroporation and mass transfer theories is presented. The outcome of the improved mathematical model can predict the outcome of electrochemotherapy based on the drug physical properties. The design of current microfluidic devices can be optimize for the consideration of the electroosmosis effect on drug uptake rate. We have simulated the novel microfluidic device to study the effect of simultaneous electrokinetics phenomena on cell lysis and uptake rate of the macromolecules by the COMSOL Multiphysics simulation.

498542 Nvu-on-a-Chip: Optimizing Brain Endothelial Cell Culture for Microfluidic Modeling of the Nvu

Abstract:

Organ-on-a-chip devices are an emerging class of *in vitro* models that combine microfabrication and spectroscopic techniques with cell culture to study organ physiology. We are developing an organ-on-chip model of the neurovascular unit (NVU), NVU-on-a-chip, which analyzes real-time NVU dynamics in a controlled microenvironment. NVU-on-chip incorporates human brain endothelial cells (hBECs), astrocytes and neuronal cell lines to mimic physiological phenomena of the neurovascular unit (NVU). A key component of the NVU is a selectively permeable layer of endothelial tissue, the blood-brain barrier (BBB), which prevents passage of most small particle nanotherapeutics from the bloodstream into the brain. We have optimized brain endothelial cell culture in a monoculture, microfluidic device for future incorporation with NVU tissue. Endothelial cells are seeded into a microfluidic organ-on-a-chip device containing a single microfluidic channel and two electrode interfaces; tissue integrity is analyzed with electrical impedance spectroscopy (EIS). The device contains a pair of (5 nm Titanium, 25 nm Gold) interdigitated electrodes that form the top and bottom layers. It contains a 100-300 μm thin PDMS channel (500 × 18000 μm) between the two electrodes which complete the microfluidic device. The PDMS channel is necessary to facilitate gas exchange to the cells in the microfluidic channels. Cell death is observed if other materials like laser cut PMMA channels are used. The top and bottom electrode/channel pairs sandwich an extracellular matrix (ECM) coated Transwell membrane seeded on one side with rat brain microvascular endothelial cells (RBMEC). The ECM coating needs to be uniform which allows the cells to grow in the microfluidic channel. The uniformity of the ECM coating is validated by using a fluorescent microscope.Cells mature for several days in a 37°C, 5% CO_2 humidified incubator post-seeding to allow the formation of barrier properties. The fabricated device is characterized using optical imaging, permeability assays, such as fluorescence microscopy, and EIS. Optical imaging confirms endothelial cell adhesion and confluency. Fluorescence microscopy signifies the presence of ZO-1, an accessory protein indicative of tight junction formation. EIS measures resistance and capacitance across the seeded endothelial membrane. A resistance value of ~1000Ω indicates a functional blood-brain barrier, while lower values implicate a compromised or 'open' BBB. EIS measurements are advantageous because they provide real-time capacitance and resistance measurements of transient BBB activities, such as permeability changes. Additionally, EIS capacitance data distinguishes transcellular resistance from paracellular resistance. This novel approach provides insight to transcellular BBB kinetics as well as paracellular (tight junction) kinetics. The device will be incorporated into a more sophisticated NVU-on-a-chip. NVU-on-a-chip will be used to characterize the interaction and mechanistic pathway for drug-loaded nanoparticles.

498993 Electrohydrodynamic Scaling Laws Analysis in a Microfluidic Isodep Device

Abstract:

Electrohydrodynamic scaling laws analysis in a microfluidic IsoDEP device

M.Z. Rashed[1], K.C. Grome[1], S.P. Hendricks[2], S.J. Williams[1]

1. University of Louisville, Louisville, Kentucky USA

2. Murray State University, Murray, Kentucky USA

Dielectrophoresis (DEP) is the phenomenon in which a particle, such as a living cell, is moved by the interaction between a non-uniform electric field and its induced polarization. Isomotive dielectrophoresis (isoDEP) is a cell analysis and characterization technique that uniquely utilizes a constant gradient field-squared (*Erms2)* resulting in a uniform DEP force. The resultant constant (isomotive) particle translational velocity that can be tracked using particle tracking velocimetry (PIV) software to extract the cell/particle dielectric properties. Inspired by initial analysis by Herbert Pohl, we have developed modified electrode geometry for isoDEP. Fabrication of extruded electrodes is straightforward via microfabrication methods (DRIE of conductive wafers) or sub-millimeter machining. A sample is injected and flow is halted before field activation. Digital images will extract particle size and, due a constant *Erms2* , the only unknown for each particle is Re[f_{CM}] The field is applied and Re[f_{CM}] is extracted through particle tracking. The particle's velocity will change as the AC frequency is swept over a specified range to obtain a comprehensive Re[f_{CM}] spectrum. Through simultaneous particle tracking such spectra are obtained for every particle in the imaging area, enabling parallel analysis of cells. IsoDEP can extract the dielectric properties of each cell (ex: membrane capacitance) – these properties directly correlate to the cell physiology. Any unwanted flow will disrupt the trajectory of the particles and compromise their analysis. To that end, we have conducted an electrohydrodynamic study and scaling law analysis to reduce electrothermal hydrodynamics in an isoDEP device. Numerical simulations (COMSOL Multiphysics) are in good agreement with experimental measurements via micro-PIV. In addition to experimental results, current and future IsoDEP platform designs will be shared.

References

[1] H.A. Pohl, *Dielectrophoresis: The Behavior of Neutral Matter in Nonuniform Electric Fields*, Cambridge; New York: Cambridge University Press (1978).

[2] Allen, D. J., Accolla, R. P. and Williams, S. J. (2017), Isomotive dielectrophoresis for parallel analysis of individual particles. ELECTROPHORESIS. doi:10.1002/elps.201600517.

500788 Fundamentals, Calibration and Preliminary Results Using the DSC Technique for Hydrogel Thermoporometry

Abstract:

Hydrogels are cross-linked polymeric networks. They can be used in many biomedical applications such as drug delivery, tissue engineering and bio-separations. Characterization of the hydrogel pore structure is very important in order to select the appropriate gel type for a particular application. *Thermoporometry* is a powerful technique that is may be used to study the structure of hydrogels. Both *pore size and pore-size distribution* may be measured by using an instrument called the *differential scanning calorimetry* (DSC). This technique relies on the melting or freezing temperature depression of the water confined in a pore. Thermoporometry by DSC is a unique method for characterizing gel networks. Since there is no general procedure for measuring pore-size, further investigation is required to establish guidelines for the analysis of specific materials. Although some formulas have been developed, choosing the right equation for calculating pore size is still challenging. Many of the parameters incorporated in these equations are specific to the gel type (e.g. the natural of the porous material, the range of the pore size tested, and the probe liquid used). For this reason, it is important to develop an equation for the specific pore size of the hydrogel of interest.

In this work, the DSC has been well-calibrated with high purity mercury. Two gel types will be characterized: regular (non-templated) gels and nanotemplated hydrogels with a modified pore structure. The results will then be analyzed in order to study the effects of the size, geometry and shape of the templating agent on hydrogel performance. Details about the implementation of the DSC Technique and discussion of the preliminary results will be included.

509909 Validation of A Novel Algorithmic Approach To Solve The Poisson-Boltzmann Equations In Electrokinetics

Abstract:

Many exploratory studies rely on the ability to make accurate estimations of the variable under study. In electrokinetics applications, appropriate prediction of electro-osmotic velocity profiles is crucial to the study of electrokinetic processes, such as electro-assisted drug delivery, micro-electrophoretic separations, soil remediation, and material processing. All these applications require the mathematical solution of the complete Poisson Boltzmann (P-B) equation for the systems under study. This equation is a complex mathematical expression characterized as a secondary nonlinear ordinary differential equation. Not long ago, Arce and Oyanader* proposed a novel and simpler solution of the P-B equation using a recursive function, $Æ'_{AO}$. In this work, the previous contribution of Arce and Oyanader will be revisited using a new, more analytical approach, but validating the simple predictor-corrector method developed and introduced by Arce and Oyanader as an alternative tool for nonlinear phenomena problems.

This validation seeks to persuade practitioners from different discipline areas, such as hydrodynamics, electrostatics, and mass and heat transfer that the recursive function, $Æ'_{AO}$, is of simple and practical use. In particular, this novel approach has proven to be an exceptional tool in modeling the electrical field for applications of interest such as the separation of a mixture of macromolecules and the removal of contaminants on soil cleaning processes. Several examples will illustrate the benefits of the methodology. Comparisons to numerical solutions are also included.

(*) Oyanader, M., Arce, P., â€œA New and Simpler Approach for the Solution of the Electrostatic Potential Differential Equation. Enhanced Solution for Planar, Cylindrical and Annular Geometries,â€ *Journal of Colloid and Interface Science*, 2005, 284, 315.

509910 Mathematical Analysis of Bone Remodeling under Influence of Electrical Field

Abstract:

Bone remodeling, defined as the replacement of bone tissue, is a process consisting of two basic mechanisms, bone resorption and bone formation. It has been found that bone formation can be influenced by the presence of an electrical field. If this finding continues to test true, applications of electrical fields may lead to novel treatments in this area of study, including bone fractures, transplant, and osteoporosis.

Numerous mathematical models have been proposed to study the bone remodeling process influenced by different types of precursors, hormones and vitamins for example. Unfortunately, none of them have analyzed the effects of electrical fields. The main focus of this contribution is to proposed a mathematical model for net bone formation and to analyze theoretically and numerically the effects of electrical field on osteoblast cells associated with bone production.

The proposed model was conceived following a second order bio-kinetic model limited by precursor availability and controlled by endogenous decay. In particular, osteoblast cellular specific growth rate was modeled using an Arrhenius type function, electrical field dependent.

Preliminary results indicates that electrical fields have a significant impact on bone formation. Increased levels of osteoblast cells are predicted when different order of magnitude of the electrical field are imposed on the system. These results provide important insights into the development of new treatment protocols for patients with bone injuries, diseases, and/or implants.

510291 Analysis of Lipemia Levels from Human Blood Samples Using Microchips

Abstract:

Lipemia is a transient status of high lipid levels in the blood that adversely affects health, donated blood quality for diagnostic tests/other. This project aims to develop a microchip device to detect lipemia with high speed, small sample, high sensitivity, and good stability analyze lipid levels in different bio-fluids. Four measurements were compared: (1) plasma electrical characterizations, (2) RBCs dielectrophoresis, (3) RBCs lysis, and (4) plasma UV-vis absorbance. UV-Vis absorbance results effectively detected free lipids level in blood of a healthy young donor for two cases: fasting for 10-12 hours, and post-meal 2-3 hours after a high sugar and fatty meal. Plasma UV-Vis absorbance results revealed substantial spectral differences between the cases. The overarching goal for this project is to detect lipemia states using a single drop of blood, which would enable nutritional and CVD screenings for medical interventions.

Unamplified and Sensitive DNA Sensor for MRSA Detection by Capacitive Sensing and Low Voltage AC Dielectrophoresis

Rania Oueslati[a], Jiangang Chen[b], Jayne Wu[a] *

[a]Department of Electrical Engineering and Computer Science, [b]Department of Public Health,

The University of Tennessee, Knoxville.

Introduction:

Methicillin-resistant Staphylococcus aureus (MRSA) is one of the most prevalent causes of nosocomial and community-acquired infections with worst clinic consequences. MRSA-colonized patients face significantly higher mortality and morbidity rates than patients infected with methicillin-susceptible S aureus. Therefore, fast and reliable methods for detection of MRSA are warranted to limit the spread of infection. This work presents a low-cost, highly sensitive, and specific DNA sensor as a screening tool for rapid detection of MRSA. The use of alternating current electrokinetics (ACEK) speeds up the hybridization of DNA molecules and improves the detection sensitivity. This capacitive sensor is highly sensitive (aM), rapid (within 30s), easy to use (no complex steps or signal amplification) and low cost (less than 10 cents).

Mechanism:

The sensing mechanism for our bacterium DNA detection is based on alternating current (AC) electrokinetics enhanced capacitive sensing. A special AC measuring signal is applied between a pair of functionalized interdigitated microelectrodes (IDMEs). Innovation includes (1) low voltage dielectrophoresis (DEP) enrichment of DNA molecules for rapid hybridization and high sensitivity; (2) Use of interfacial capacitance (Cint) as the indicator of specific binding, and the direct measurement of Cint for robust operation and minimal matrix effects; (3) Simultaneous execution of (1) and (2) during assay to achieve a single step operation. Experiments:

Experiments:

<u>*Electrodes and probe density:*</u>

Two types of electrode materials were tested, electroplated gold on printed-circuit (PCB)-based IDMEs and aluminum IDMEs. The plated gold electrodes were shown to immobilize more probe than aluminum electrodes, and yielded larger responses. However, their capacitive responses cannot distinguish between positive and negative samples. Aluminum IDME had lower probe density, which allows target DNAs to approach the IDME to exhibit a decreasing capacitance change, while non-targets lead to an increasing capacitance change. So aluminum IDMEs were adopted.

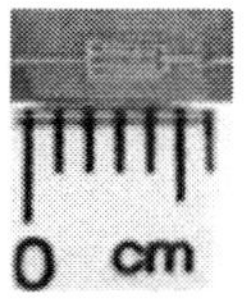

Electrode1: printed-circuit based IDME Electrode2: Aluminum IDME

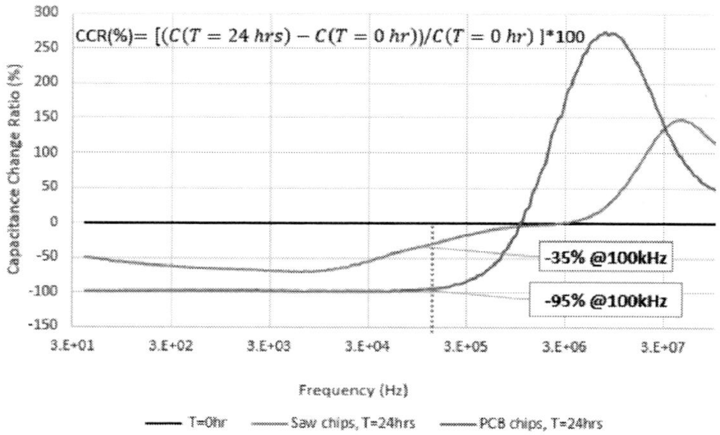

Fig1: Capacitance change Ratio after probe incubation on SAW

(Blue) and PCB (orange) electrodes

Voltage optimization:

The measuring AC signal was carefully chosen. It should induce sufficiently strong DEP attraction of DNA molecules and yet not so strong for the detection to be non-specific. 7 mV, 10 mV and 15 mV was tested, and 10 mV was found to be the best of the three.

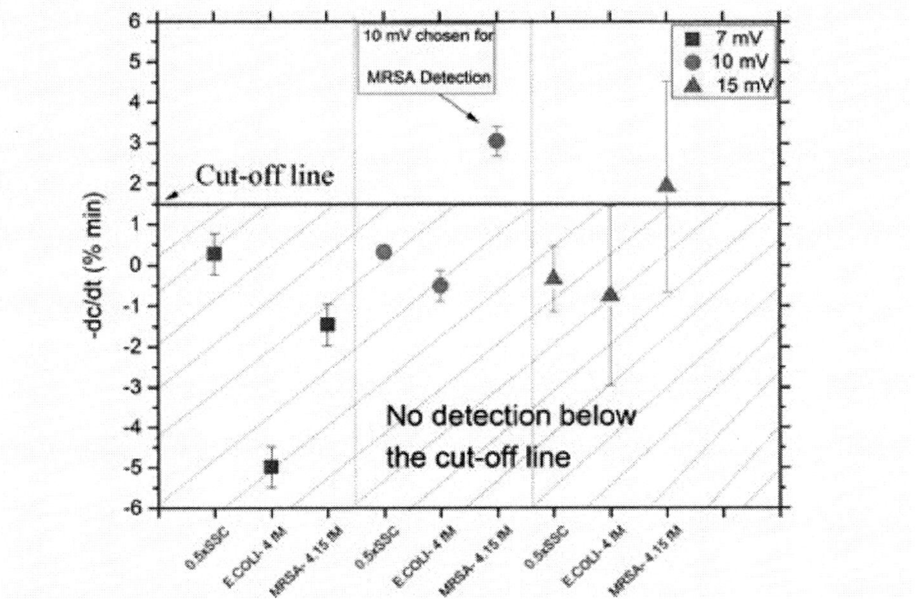

Fig2: MRSA Detection: voltage optimization

Results and Conclusions:

For plated Au PCB chips, capacitive responses cannot distinguish between positive and negative samples. The limit of detection is defined as three times the standard deviations from the background response.

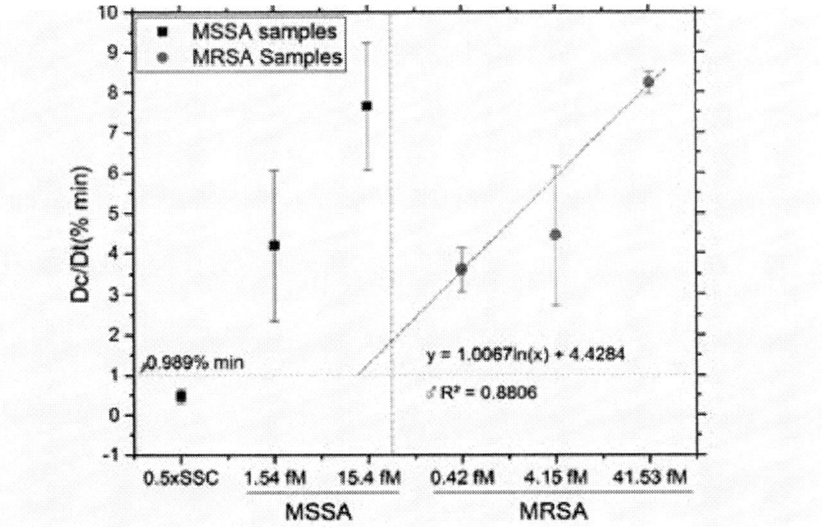

Fig3: MRSA Detection on PCB electrodes incubated with MEC probe.

The probe EB (5’-TTACAGAGTTAACTGTTACC-3’) is designed to detect femB gene, common in both MRSA and MSSA (methicillin susceptible S. aureus). The sensor, as shown in Fig4a, responds similarly to MRSA and MSSA when targeting their common gene.

The MEC probe is designed to detect femB gene, common in both MRSA and MSSA (methicillin susceptible S. aureus). The sensor, as shown in Fig4b, responds similarly to MRSA and MSSA when targeting their common gene.

(a)

(b)

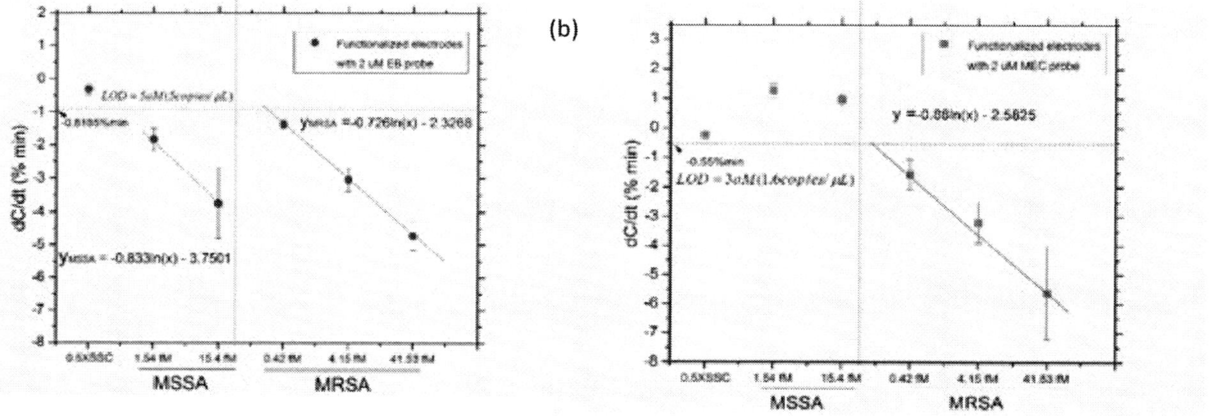

Fig4: MRSA Detection on saw Chips incubated with: (a) EB probe and (b) MEC probe

This sensor is capable of direct DNA detection with a response time of 30 s. The LOD was calculated to be 8 pg/mL in standard buffer. The dose response was found to be, where x is DNA concentration. The sensor is specific against MRSA. The detection does not need any signal amplification, which considerably simplifies the sensor operation and makes it highly suitable for point of care disease diagnosis

510767 Electro-hydrodynamics of Soft Liquid Metals at Low Voltages

Abstract:

This work characterizes the behavior of a eutectic alloy of gallium and indium (75% Ga, 25% In, by weight, 'EGaIn') in response to electric fields. The metal is a liquid at room temperature (M.P., 15.5 °C) and exhibits low toxicity. These fluidic metals may be injected into microfluidic systems, fibers, and capillary networks to form soft electronic devices that are soft and compliant. Once injected, the metal remains in its place because of the adhesive nature of its thin native oxide. Preventing the oxide adhesion within microchannels enables reversible actuation of EGaIn. Actuating liquid metals may be useful for soft actuators, reconfigurable optical displays, frequency tunable antennas, and other opto-fluidic technologies.

In this work, we utilize low voltages (<2 V) to reversibly move droplets of EGaIn through microchannels. Pre-wetting the channels with an aqueous solution prior to injecting the metal prevents oxide adhesion; the water forms an interfacial 'slip-layer' the metal and channel wall. Thereafter, an applied electric field (~10-20 V/m) actuates the liquid metal by establishing a gradient of surface tension; this effect is known as continuous electrowetting (CEW). Although CEW has been utilized before with mercury, which is toxic, the adhesive nature of the Ga oxide complicates CEW behavior. This work utilizes optical microscopy, cyclic voltammetry, and electrochemical impedance spectroscopy (EIS) to characterize electro-hydrodynamics of the system under a variety of conditions. Specifically, we elucidate the influence of electrolyte (i.e., composition, pH, and viscosity) on the metal-electrolyte interface. In addition, we compare electro-hydrodynamic behavior of EGaIn with and without the presence of an oxide 'skin.' Finally, this work explores novel microfabrication strategies to design interfaces that prevent oxide adhesion.

Rapid and Sensitive On-Site Serodiagnosis of Pseudorabies by AC Electrokinetics-enhanced Capacitive Sensing

Cheng Cheng[a], Rania Oueslati[a], Shigetoshi Eda[b], Jayne Wu[a]

[a]Department of Electrical Engineering and Computer Science, [b]Department of Forestry, Wildlife and Fisheries

The University of Tennessee, Knoxville, TN 37996, USA

Motivation:

Pseudorabies, also called Aujeszky's disease, is caused by infection of animals with pseudorabies virus (PRV). Since early 1980, pseudorabies spread worldwide and caused significant economic loss to swine industry. In the U.S., the annual cost for eradication of pseudorabies is estimated to be $30 million including $17 million for vaccination efforts. Serological surveys in the U.S. for pseudorabies in feral hog populations revealed up to 61% of tested animals (average 21.5% with 95% confidence interval of 13.4-29.8%) were positive for PRV infection, showing widespread of the disease in the animal populations [1]. Therefore, detection and control of PRV infections both in domestic and wild pigs (feral hogs) are critically important to prevent outbreak of the disease.

Diagnosis of pseudorabies is currently conducted mainly by serology, such as enzyme-linked immunosorbent assay (ELISA), serum neutralization test (SNT) and latex agglutination test (LAT) [2]. Although the serological tests are reliable, they do not differentiate naturally infected animals from vaccinated animals. Also, they cannot detect infection until adaptive immunity is developed enough for the animals to produce antibodies against the virus, which usually takes seven days. Fluorescent antibody tissue section test (FATS) can be used to confirm PRV infection by detecting virus in a matter of hours but requires tissue samples from suspected animals. Virus isolation (VI) also detects virus particles but it requires mammalian cells to propagate the virus. These tests are conducted in diagnostic laboratories and therefore suffer from long turnaround time and high cost. Therefore, there is an urgent need for a rapid, inexpensive, field-deployable diagnostic test for this disease.

***Mechanism*:**

This work is based on alternating current electrokinetics (ACEK) capacitive sensing method. The major innovation with this sensing technology is its ability to simultaneously generate microfluidic ACEK effects with the sensing electrodes and to interrogate directly the sensor's interfacial capacitance. It is based on measuring the interfacial capacitance (Cint) on functionalized microelectrodes with an AC signal capable of inducing ACEK effects. With binding process continuing, the thickness of interfacial layer, dint, keeps increasing, which leads to a decrease of Cint. By monitoring the change rate of Cint, specific binding of analyte with probe molecules can be detected. The change rate in interfacial capacitance is tracked and used as an indicator of probe-analyte binding.

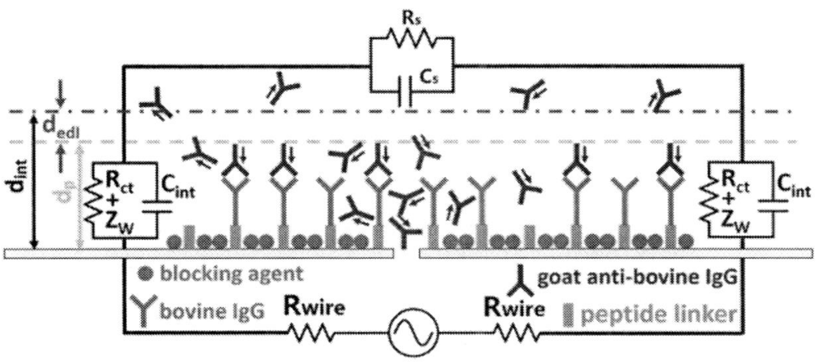

Fig. 1. Changes in EDL layer of the electrodes at different stages of sensor functionalization and Pseudorabies binding.

In order to accelerate binding reaction occurred on the electrodes surface, the AC signal will also induce movement of proteins towards the electrode surface and accelerate the binding to immobilized probe molecules. Based on our prior work [3], electrodes with larger characteristic length have remarkably higher response at the same electric field strength because the electrode geometry is more amenable to ACET convection. The findings have been applied here to achieve faster and more sensitive detection. Both positive dielectrophoresis (pDEP) and AC electrothermal (ACET) effect have been induced to transport target proteins towards the sensor. Positive DEP will caused a particle to move towards high electric field region due to interactions between a particle's dipole moment and a non-uniform field, while ACET effect will generate vortices-like microflows around an electrode and transport bioparticles by convection. Since DEP force is short-ranged, concomitant generation of ACET flows will significant improve the effective range of macromolecular enrichment.

Results:

In this work, a capacitive immunosensor based on ACEK has been developed for detection of pseudorabies virus antibody in serum samples. Interdigitated electrodes with critical dimension of 100 μm have been used as the sensor. Comparing with previous work using 2 μm electrodes [4], the optimal frequency has shifted from 100 kHz to 75 kHz. This is because ACET effect becomes more important relative to DEP with larger electrodes, and capacitive response is larger at a lower frequency with ACET effect, generally speaking.

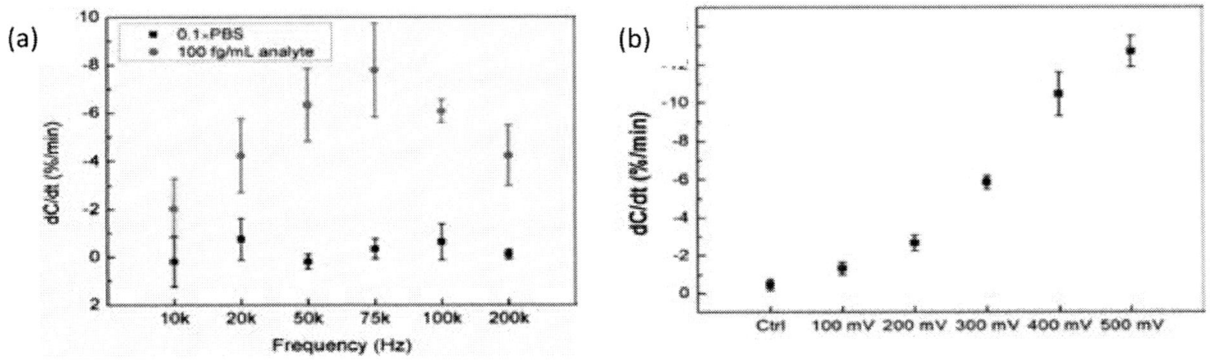

Fig. 2: The effect of the AC electric field ((a) Frequency and (b) amplitude) on Pseudorabies detection

The sensor is very sensitive with a limit of detection of 4.5 fg/mL when testing analytical samples. When testing clinical samples, using 35 serum samples from feral hogs, a diagnostic sensitivity of 91.1-% and a selectivity of 90.9-% are achieved, as confirmed by ELISA results. This sensing method has a response time of 30 seconds and requires minimum pretreatment without washing process, which make it a promising method for on-site disease detection and diagnosis.

References

[1] L. E. Pomeranz, A. E. Reynolds, and C. J. Hengartner, "Molecular Biology of Pseudorabies Virus: Impact on Neurovirology and Veterinary Medicine," Microbiol. Mol. Biol. Rev., vol. 69, no. 3, pp. 462–500, Sep. 2005.

[2] T. Müller et al., "Pseudorabies virus in wild swine: a global perspective," Arch. Virol., vol. 156, no. 10, pp. 1691–1705, Oct. 2011.

[3] H. Cui et al., "Rapid and sensitive detection of small biomolecule by capacitive sensing and low field AC electrothermal effect," Sens. Actuators B Chem., vol. 226, pp. 245–253, Apr. 2016.

510764 Chemo-Electro-Thermotherapy in Capillary Systems: Simplify Model and Simulation

Abstract:

The use of electrical fields in medicine include a long range of practical applications for diagnosis and treatment of recurrent contemporary illnesses. A few examples of these applications are biological imaging, neurology, iontophoresis, drug delivery, and the most recently Chemo-Electro-Thermotherapy for cancer therapy. In a previous contribution, these authors identified a framework for undergraduate research on the subject[1]. This exploratory research included multiple categories: Cancer and the major effects of Angiogenesis, Tumor Growth and Electrical Fields. From a mathematical modeling approach, it was concluded that further effort was needed in five distinctive areas to conduct effective product design and development within the field. In this current work, at least two areas of efforts are reported and further developed to analyze the role of electrical fields, Joule heating, interstitial pressure and systemic or intramural drug injection. The main contribution of this research is demonstrating the effects of the order magnitude of the applied electrical field and drug concentration gradient across a capillary bed on malignant growths. The results will show temperature gradients in the arterial and tissue domains as a consequence of the applied number of voltages. Preliminary report on the cell count as well as drug uptake will be also shown.

1. A Framework for Undergraduate Research on Chemo-Electro-Thermotherapy, ***AIChE Annual Meeting***: American Electrophoresis Society, San Francisco, CA, Nov. 14, 2016. **F. Smallwood**, Chirstopher Fernandes, S. Oyanader, M. Oyanader.

510763 Electro-Aided Peritoneal Dialysis: A Fundamental and Modeling Analysis Approach

Abstract:

Renal replacement therapy can be conducted by using processes including Hemodialysis, Peritoneal dialysis, and a few others. Hemodialysis is not very popular among patients due to the possible yet debilitating side effects, which may involve low blood pressure, headache, tiredness, skin irradiation, and muscle cramps. Peritoneal dialysis, the alternative, is considered a much more patient friendly therapy when compared hemodialysis. Although, the most important drawback in Peritoneal dialysis is the low permeability of the peritoneal membrane, which makes the treatment significantly slower, often requiring between three to four batches of treatment per day. Membrane permeability can be modified by imposing a low value electrical field adjacent to the peritoneal wall. This soft tissue has been observed to be susceptible to electrical impulse and therefore allowing this characteristic to be exploited. This contribution analyzes the fundamental principles involved in permeability manipulation, and a mathematical model has been proposed to further study the phenomenon. The influence of different order of magnitude of the electrical fields has been studied to assess potential enhancement of the treatment in time duration and number of batches per day. The influence of blood shear thinning characteristics have also been investigated and illustrative results will be presented in the analysis.

511394 DNA Gel Electrophoresis via Entropic Trapping: Insights From Monte Carlo Simulations

Abstract:

The ability to precisely manipulate transport of charged macromolecules (DNA, proteins) in nanoporous surroundings plays a critical role in the design of microfluidic electrophoresis systems. Reptation is the predominant transport mechanism in most hydrogel-based separation matrices, where the characteristic macromolecular size scale is larger than that of the surrounding pores. DNA electrophoresis in the reptation regime yields a scaling of mobility with the inverse of macromolecular size. Alternatively, when the characteristic macromolecular size is the same as that of the surrounding pores, transport occurs via an entropic trapping mechanism. In contrast to reptation, transport in this regime occurs via discrete hops between larger pores joined by narrow interconnecting spaces, leading to a stronger size dependence of electrophoretic mobility that is desirable for enhanced separation performance.

We have previously shown how application of a time varying electric field driving force can enable the activated hopping process to become synchronized, leading to a resonance state that can be exploited both to obtain improved separation performance and to extract nanoscale physical parameters of the macromolecules (contour length, persistence length). In these initial studies, our transport model was focused on the activated process by which molecules enter the narrow space between adjacent large pores (activation time), but assumed a simplified representation of transit through the confined interconnecting region (migration time). Specifically, migration was described in terms of a simple function of the macromoleculeâ€™s radius of gyration and the applied electric field. Although this representation is consistent with previous frameworks that have been developed to describe electrophoretic entropic trapping, it does not realistically capture the hydrodynamic forces and conformational changes that occur during migration through the confined interconnecting pore spaces.

Here we apply a more realistic representation of migration through the nano-confined interconnecting spaces linking adjacent large pores. Our starting point is a model is based on a Monte Carlo algorithm developed by Slater, et. al. that directly incorporates frictional and entropic forces. The portion of the macromolecule located outside of the pore space experiences frictional interactions with the surrounding fluid, while the portion inside the pore experiences drag due to spatial confinement in the narrow interconnecting space. Entropic forces are related to conformation differences between the portion of the macromolecule at the entrance and exit of the pore space. Embedding each of these effects into our model enables the size dependence of electrophoretic mobility to be computed by integration over the pore size distribution of the hydrogel matrix. The resulting predictions are validated by comparison with experimental data. These new insights can suggest optimal operating conditions (electric field amplitude and frequency) and nanoporous gel matrix architectures (pore size distribution) to deliver enhanced separations in microfluidic electrophoresis systems.

511396 Dielectrophoretic Lipid Content Differentiation in *Neochloris Oleoabundans* for Biomass Harvesting Optimization

Abstract:

The dielectrophoretic behavior of Neochloris oleoabundans microalgae with heterogeneous cytoplasmic lipid content was characterized by dielectrophoresis using carbon interdigitated electrodes. For this purpose, two samples of microalgae were cultured: one under nitrogen-replete (N+) and the other under nitrogen-deplete (N-) conditions to allow low and high lipid growth, respectively. Cell population was monitored by spectrophotometry and the difference in lipid content among the samples was determined by Nile red fluorescence. The microfluidic device consisted of a carbon castellated microelectrode array fabricated using the Carbon-MEMS processâ€"photolithographic patterning of a carbon precursor, followed by pyrolysisâ€"bonded to a poly(dimethylsiloxane) microchannel. A Finite element model was developed to determine the electric field distribution across the microchannel and dielectrophoretic trapping zones. Experimental tests were carried out on wide frequency window (from 100 kHz to 30 MHz) at a fixed amplitude of 7 V_{PP}. Results showed a significant difference between the dielectrophoretic behavior of N+ and N- cells at low frequencies (100-800 kHz), whereas a weak response for mid and high frequencies (1-30 MHz). These results suggest that pyrolyzed carbon, obtained from a low-cost and straightforward fabrication process, is an attractive electrode material for microalgae dielectrophoretic characterization and isolation, based on their lipid content.

511398 Toward the Design of a Multi-Module Fluidic Device for the Simultaneous Detection of Lyme Disease and Babesiosis

Abstract:

In our previous work on the dielectrophoretic separation of Babesia-infected erythrocytes from its mixture with healthy ones, we leveraged the pathogenesis and clinical courses of both Babesia and Plasmodium to theorize their interchangeable electrophysiological properties, which were used to simulate the trajectory of the Babesia-infected erythrocytes in a 1.4mm long fluidic device. When the Trends in Parasitology released one of its issues in 2015, it clearly reported some statistical events which showed that 40% of patients with Babesia-infected erythrocytes are not mutually exclusive with patients who were infected with Borrelia-the main cause of Lyme disease. This sets the diagnostic activities of these pathogens out for both deterministic and stochastic considerations. A foundational necessity is the characterization of these pathogens. This work, therefore, focuses on the characterization of Babesia-infected erythrocytes in a perpendicularly arranged electrode pair of 75 μm inter-electrode spacing sizably contained in a PDMS-constructed micro well. Trapping and streaming activities at 8Vpp and various frequencies, which led to the determination of the cross-over frequencies at various buffer conductivities were observed and recorded. Curve fitting procedure for these data (crossover frequency v. buffer conductivity) led to the estimation of both shell and cytoplasmic electrical parameters. Ongoing work will characterize Borrelia.

510559 Dielectrophoretic Response of Condensed DNA Clusters in AC Fields

Abstract:

Insulator-based dielectrophoresis (iDEP) has been applied in analytical research as a method to separate and fractionate DNA in lab-on-a-chip devices. The development of similar continuous-flow, size-based separation techniques for DNA Next-Generation Sequencing (NGS) could advance the speed, efficiency, and accuracy of in situ analysis of genomic/biological data. When low-frequency and high-amplitude AC potentials are applied to homogeneously disperse Î»-DNA, reversible segregated clusters are observed in high frame rate imaging. However, the fundamental physics of such cluster phenomena are poorly understood and limited studies have been published referencing the aggregation formation realized under these conditions. This recently observed clustering phenomenon has gained the attention of colloidal physicists who wish to understand the physical parameters that influence clustering of Î»-DNA. We believe applied low-frequency potentials generate this clustering behavior, which influences the dielectrophoretic response of DNA molecules. As a result, this phenomenon could be used to improve DNA sample preparation and pre-concentration using microdevices for future NGS applications. Research investigating cluster formation through modeling and computer simulations has shown that this phenomenon may be due to electrohydrodynamic instability including by not limited to dipole-dipole interactions between molecules. However, these models often address simpler DNA systems and fail to predict the unexpected switch we observe from positive to negative DEP in DNA clustering. Clustering behavior has the potential to enhance the efficiency of DNA fractionation by exploiting strong AC electric fields to induce condensation and separation of DNA molecules, thus it is important that we understand the physics behind this phenomenon. Our recent work examines DNA cluster formation and migration characteristics using various electric field parameters in order to quantify the movement of DNA particles in solution. Preliminary evidence suggests that the electrokinetic mobility of DNA clusters changes negligibly with changes in frequency. Various amplitudes and frequencies of applied potential were examined to determine the extent of correlation in DNA clustering. DNA clustering was largely observed with frequencies ranging from 10 to 100 Hz and electric field strengths above 800 V/cm. Similar electric field parameters have produced some of the highest-observed sorting efficiencies in an iDEP constriction sorter for DNA molecules ranging in size from 10 to 50 kbp. Future work will aim to enhance techniques capable of automatically tracking DNA clusters throughout high-frame rate imaging to further understand clustering and transport mechanisms. Such techniques will significantly reduce biosensing and bioanalytical processing times.

503471 Physical Properties of Bioparticles and High Resolution Separations with Dielectrophoresis

Abstract:

Cells have a much more complex electric signature in the presence of an applied external field than is typically discussed in the introduction to most papers on electrokinetic manipulations of cells. This talk will focus on cataloging and characterizing theoretical and experimental physical features and the forces associated with them for cells undergoing electrokinetic movements. While much is known about the behavior of cells in electric fields, there also remains a good deal that is poorly characterized or completely unknown. The physical features are very important to understand as the resolution of cell separations improves, allowing unequivocal identification based on electrokinetic measurements alone. In addition to discussing the physical features of cells, new device designs which promise higher resolution will be presented.

495472 Isomotive Dielectrophoresis Based Characterization of Chlamydomonas Cells

Abstract:

Isomotive dielectrophoresis based characterization of Chlamydomonas cells

M.Z. Rashed[1], K.C. Grome[1], S.P. Hendricks[2], S.J. Williams[1]

1. University of Louisville, Louisville, Kentucky USA

2. Murray State University, Murray, Kentucky USA

Dielectrophoresis (DEP) is the phenomenon in which a particle, such as a living cell, is moved by the interaction between a non-uniform electric field and its induced polarization. Isomotive dielectrophoresis (isoDEP) is a cell analysis and characterization technique that uniquely utilizes a constant gradient field-squared (Erms2) resulting in a uniform DEP force. The resultant constant (isomotive) particle translational velocity that can be tracked using particle tracking velocimetry (PIV) software to extract the cell/particle dielectric properties. The results obtained in this work are expected to significantly enhance the research on characterization and differentiation of biological cells.

The DEP force of a homogeneous spherical particle is

FDEP =2pema3RefCM Erms2

fCM=ec-em*ec*+2em**

where *ec**, *em** are the complex permittivity of the fluid medium and cell defined as e^*=e- (s/ω) j, ***a*** is the particle radius, f_{CM} is the Clausius-Mossotti factor, ***E*** is the electric field.

In this work we have achieved a cell characteristic analysis for Chlamydomonas cells (diameter:10-15 μm, Ph=8.5, suspended in a cultural medium (s_m = 19.3 mS/m) by using a prototype isoDEP device **[1]** that's fabricated using deep reactive ion etching (DRIE) of a conductive silicon wafer (1-10 O-cm) whose patterned features served as electrodes and microchannel sidewalls simultaneously that are fabricated to follow a specific curvature inherited from Pohl's**[2]** electrode design to ensure a constant (Erms2) .

The AC frequency is swept over a specified range to obtain a comprehensive f_{CM} spectrum of the Chlamydomonas cells. If Re[$\mathbf{f_{CM}}$] ,which is the real part of $\mathbf{f_{CM}}$ function, which depends on the angular frequency (ω) of the applied voltage. We noticed a negative DEP (nDEP) force at frequencies below the estimated crossover frequency of 11.6 kHz, and a positive DEP (pDEP) starting from that frequency ,where cells started to increase their translational

velocities outward the high field with increasing the frequency above the later frequency reaching the MHz range . It's noted that the cells' viability can become uncertain after the application of the AC signal (1000 V/m) for 10-12 mins **[3]** based on which requires continuous inspection of the cells from time to time while conducting the experiment – work is ongoing to assess the viability of exposed cells.

References

[1] Allen, D. J., Accolla, R. P. and Williams, S. J. (2017), Isomotive dielectrophoresis for parallel analysis of individual particles. ELECTROPHORESIS. doi:10.1002/elps.201600517.

[2] H.A. Pohl, *Dielectrophoresis: The Behavior of Neutral Matter in Nonuniform Electric Fields*, Cambridge; New York: Cambridge University Press (1978).

[3] Qian, Cheng et al. "Dielectrophoresis for Bioparticle Manipulation." Ed. Bing Yan. International Journal of Molecular Sciences 15.10 (2014): 18281–18309. PMC. Web. 16 Apr. 2017.

Characterization of Chemical Affinities and Interactions with Lipid Bilayers using Electrokinetic Techniques

By: William M. Penny & Christopher P. Palmer

UNIVERSITY OF MONTANA

Outline

- Nanodisc
- Electrokinetic Chromatography
- Indirect measurement of Hydrophobicity
- Linear Solvation Energy Relationships & Dynamic Light Scattering
- Future Directions

Nanodisc

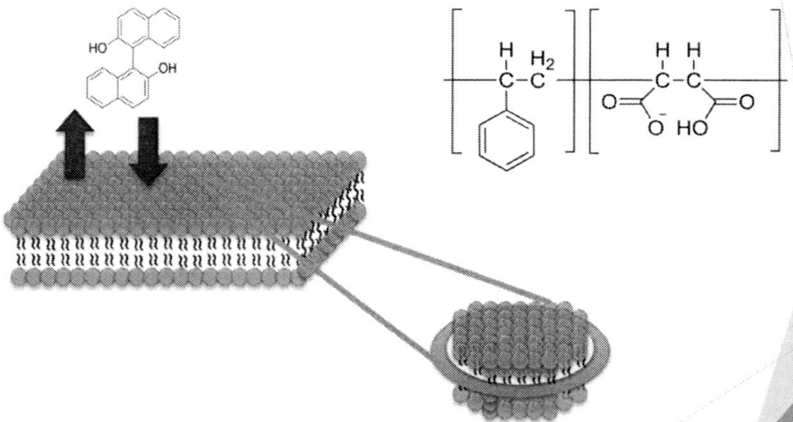

Electrokinetic Chromatography

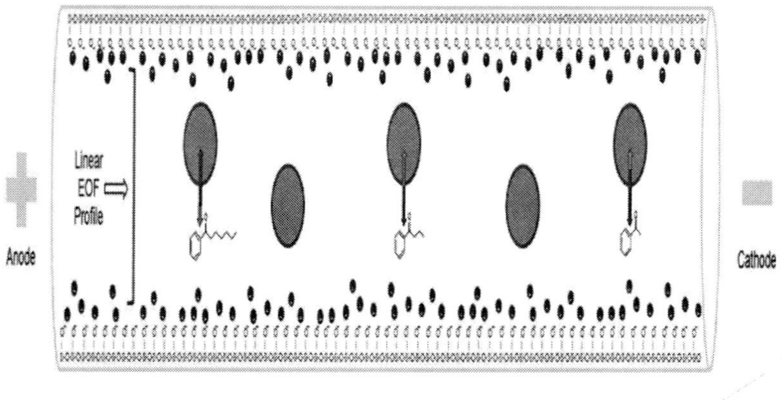

Nanodisc Composition

Xiran 30010

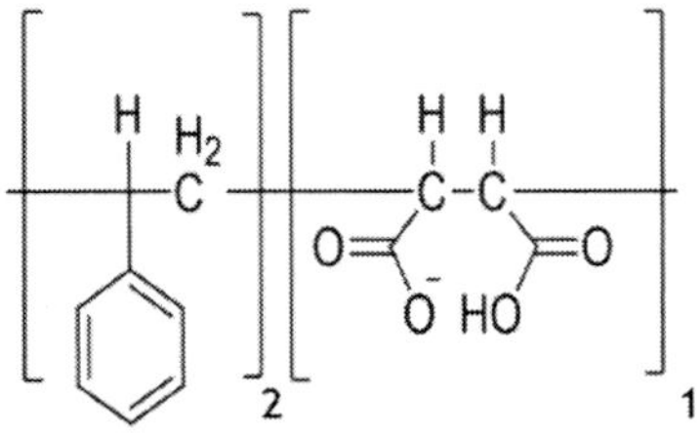

14:0 PC

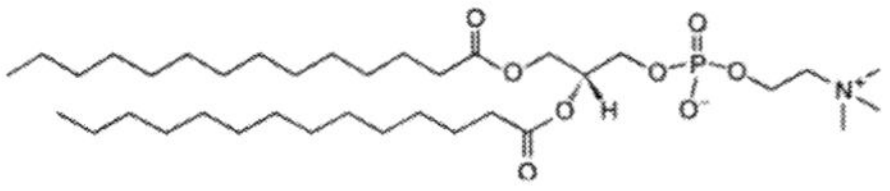

Nanodiscs contained a 0.85:1.00 copolymer: lipid ratio (w:w)

Indirect Measurements of Hydrophobicity

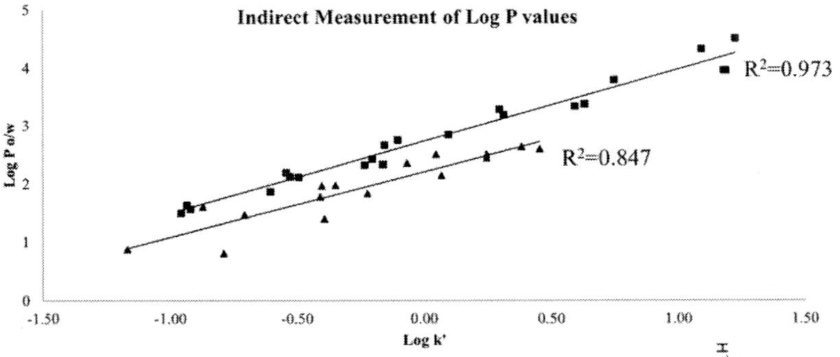

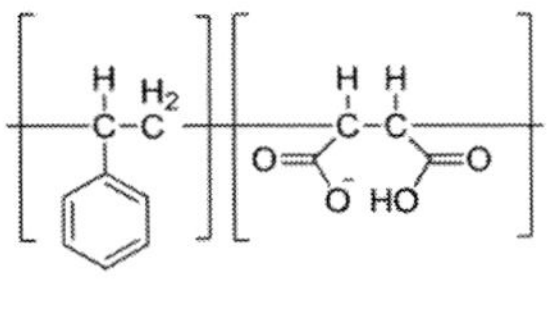

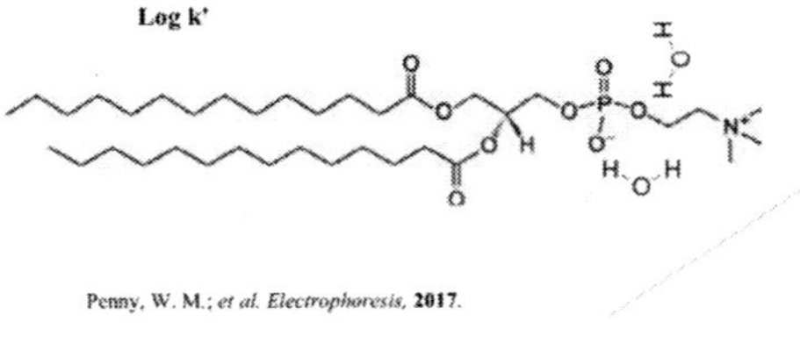

Penny, W. M.; *et al. Electrophoresis*, **2017**.

Linear Solvation Energy Relationships

- LSER analysis is used to study the free energy transfer of a solute from the background electrolyte to the nanodisc
- SP = $c + eE + sS + aA + bB + vV$
  - Excess molar refraction (E)
  - Dipolarity/polarizability (S)
  - Hydrogen bond acidity (A)
  - Hydrogen bond basicity (B)
  - Molar volume (V)

Comparison of LSER parameters based on Xiran 30010 to 14:0 PC ratio

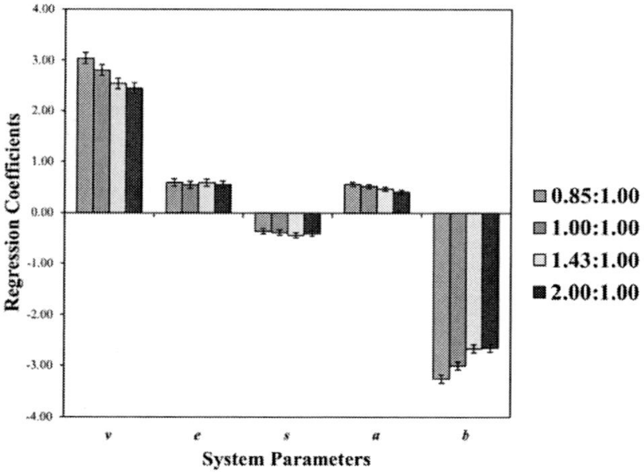

Copolymer: Lipid Ratio (w: w)	Diameter (nm)
0.85:1.00	19.3±8.34
1.00:1.00	15.6±6.48
1.43:1.00	10.3±3.14
2.00:1.00	10.4±4.16

Penny, W. M.; *et al. Electrophoresis*, In press.

Lipids Analyzed using EKC

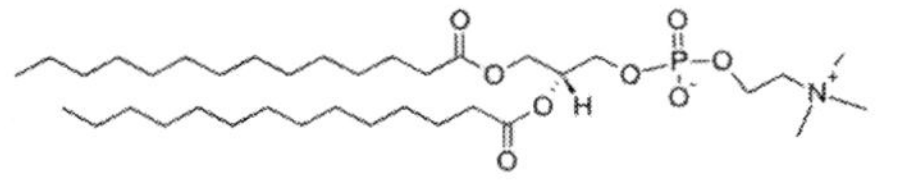

1,2-dimyristoyl-sn-glycero-3-phosphocholine (14:0 PC)

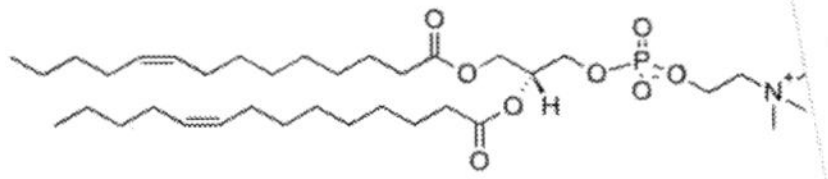

1,2-dimyristoleoyl-*sn*-glycero-3-phosphocholine (14:1 PC)

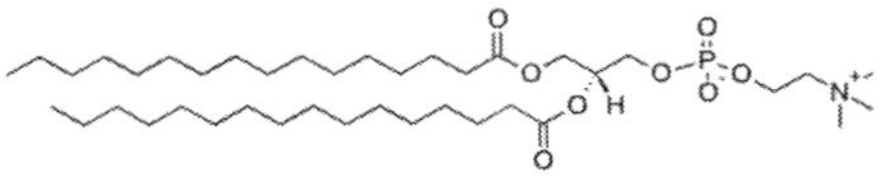

1,2-dipalmitoyl-*sn*-glycero-3-phosphocholine (16:0 PC)

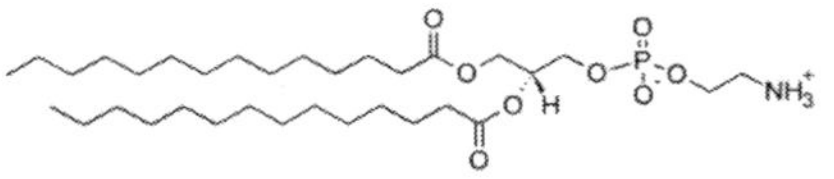

1,2-dimyristoyl-*sn*-glycero-3-phosphoethanolamine (14:0 PE)

Comparison of LSER results based on lipid composition

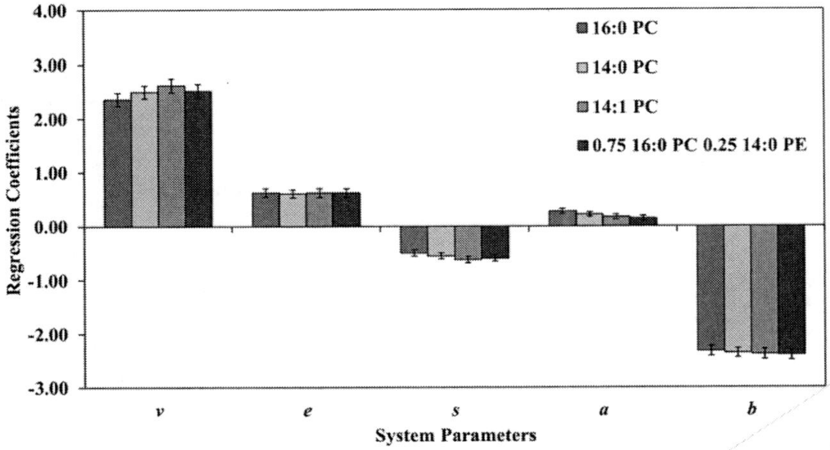

Penny, W. M.; *et al. Electrophoresis.* In press.

Comparison of lipid head chemistry

Separation of six solutes

- (1) Benzyl Alcohol
- (2) Methyl Benzoate
- (3) Ethyl Benzoate
- (4) 4-Ethylphenol
- (5) Propyl Benzoate
- (6) 4-Chlorotoluene

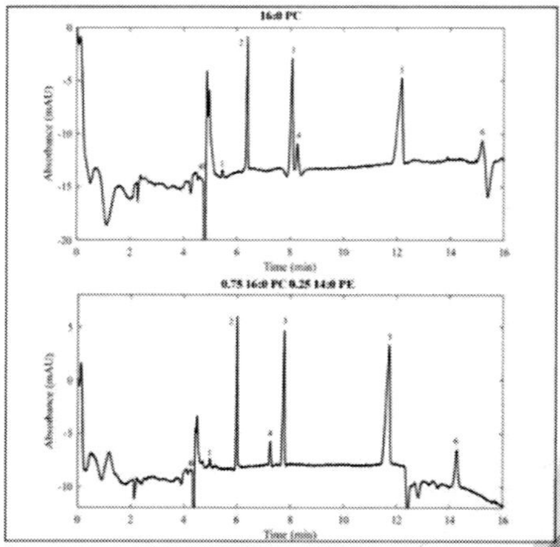

Future Directions

- Chiral separation using enantiomerically pure lipids as a chiral selector
- Measuring binding of cytochrome c to cardiolipin nanodiscs

- Penny, W. M.; Palmer, C. P. ""Determination of lipid bilayer affinities and solvation characteristics by electrokinetic chromatography using polymer-bound lipid bilayer nanodiscs."*In press, Electrophoresis*.
- Penny, W. M.; Steele, H. B.; Ross, J. B. A.; Palmer, C. P. "Phospholipid bilayer affinities and solvation characteristics by electrokinetic chromatography with a nanodisc pseudostationary phase."*Electrophoresis* **2017**, *38* (5), 738–746.*
  - ***Featured in biomedical-advances.org's New Technology section**

Acknowledgements

- Dr. Chris P. Palmer & Dr. J.B. Alexander Ross
- Harmen Steele, Jesse Hyslop, Julie McGettrick, Derek Schultz, and Hannah Wright
- Stefan Scheidelaar and J. Antoinette
- This research is supported by an NIH award to the Center for Bimolecular Structure and Dynamics at the University of Montana (P20GM103546)

MREDI Montana Research & Economic Development Initiative

Comparison of nanodiscs with different copolymer chemistries

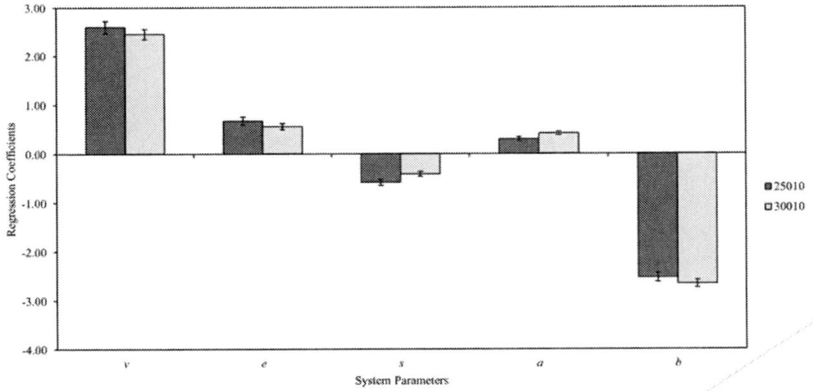

Penny, W. M.; *et al. Electrophoresis.* In press.

Comparison of liposome and nanodisc LSER analysis

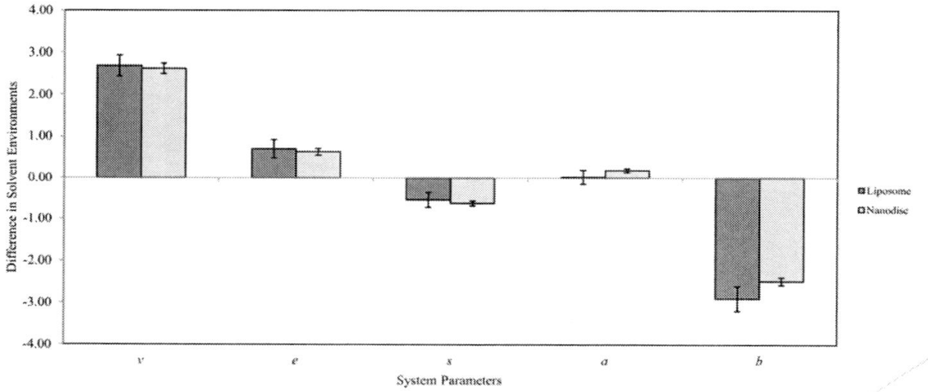

Penny, W. M.; *et al. Electrophoresis*, In press. Pascoe, R. J.; *et al. Electrophoresis* **2003**.

500790 Numerical Model for Streaming Dielectrophoresis

Abstract:

Numerical Model for Streaming Dielectrophoresis

Rucha Natu, Monsur Islam and Rodrigo Martinez-Duarte

Department of Mechanical Engineering, Clemson University, Clemson, SC, USA

Here we present initial results for developing a simulation based model for particle separation using streaming dielectrophoresis (DEP). The technique of DEP induces particle movement in the presence of non-uniform electric field, towards or away from the field. This technique has been widely used for selective manipulation and trapping of wide variety of particles and cells. Combination of DEP with a flow field is an important technique for rapid separation. Streaming DEP works with an interaction of DEP and Drag force to form concentrated streams of selected particles. In the DEP Drag force field, the dominance of DEP force gives rise to trapping of particles, whereas dominance of Drag force gives rise to flow of particles. The effective manipulation of these two forces can create concentrated streams of the particles. The variation of particle streams as an effect of these forces is studied here by implementing COMSOL simulations. The stream widths are computed for different flow rates, frequencies, and voltages in a microfluidic chip. Simulations are also carried out for different shapes of electrodes to study the streaming zones for each case. Simulation considers 1um latex particles suspended in water as media. Simulations were experimentally using a 3D carbon electrode set up.

Simulation model considers a 3D carbon electrode set up, as fabricated by use of SU-8 photolithography. [1] The electrode domain consists of an array of 80 rows with 5 electrodes in each row, with alternate polarization. Streams are analyzed for circular, diamond and triangular shaped electrodes. The flow rate is varied from 10 to 1000ul/min and the frequencies used are 10-1000 kHz. The conductivity for the media is varied from 0.0005-0.5 S/m. The analysis of streams to compute the width of concentrated streams is completed using Matlab 2014, as detailed in our previous work. [2] By varying each parameter, the average magnitude of DEP and Drag force in the domain were calculated.

The stream width was obtained for different ratios of average Drag and DEP forces. It was observed that, for every shape of electrode, the device showed streaming only is a certain range of force ratio. In the regions with ratio being higher than this range, no streams were seen, whereas for the domain with force ratio lower than the range, all particles were trapped. For a particular electrode geometry, this range remained unchanged irrespective of the voltage, flow rate and the frequency used in the simulation. The plot for force ratio and the change in stream width for triangles, circles and diamonds is shown in the Figure 1a. Figure 1b shows the comparison for streams obtained in simulation compared to experiments. This comparison is at the flow rate of 100ul/min, with voltage 14V and frequency of 50 kHz. In case of the simulation, the streams are seen to leave the electrodes at the lateral sides, whereas in the case of experiment, streams focus at the center of the electrode. This difference in behavior is due to the nature of boundary layer mesh used in the simulations and the fact that the electrodes have a larger active surface area due to surface roughness than simulated. Irrespective of these differences, the nature of the streams in both cases is very similar.

The ongoing work is develop a technique to predict the streaming behavior in the dielectrophoretic chip based on applied parameters, i.e. voltage, frequency, flow rate, media conductivity and electrode geometry. The objective of this tool is to vary the parameters to cater the trapping or streaming behavior of cells in a dielectrophoretic chip as desired.

Figure 1: a) The figure shows the width of stream obtained at different simulated voltages, frequencies, flow rates and media conductivities with respect to corresponding force ratio for different shapes of electrodes. B) Validation of simulation using experiment at the flow rate of 100 ul/min and the frequency of 50kHz.

References:

1. Martinez-Duarte R., Renaud P., and Madou M., 2011, "A novel approach to dielectrophoresis using carbon electrodes," *Electrophoresis*, **32**(17), pp. 2385–92.

2. Natu R., and Martinez-Duarte R., 2016, "Numerical Model of Streaming DEP for Stem Cell Sorting," *Micromachines*, **7**(12), p. 217.

500943 Automated Selective Cell Manipulation Using Dielectrophoresis

Abstract:

Automated Selective Cell manipulation using Dielectrophoresis

Rucha Natu, Monsur Islam, Rodrigo Martinez Duarte

Department of Mechanical Engineering, Clemson,SC

Here we present a novel technique for automated cell manipulation using a dielectrophoretic chip with motion activation using a 3d printer. Dielectrophoresis is widely used for selective cell manipulation and trapping. Due to its ability for rapid, label free cell sorting at a high efficiency, dielctrophoresis is a promising technique for sample preparation. Use of 3D electrodes has enhanced the region for action of DEP field and shows a capability of obtaining high throughput. The technique developed by our group uses the advantage of 3D carbon electrodes as a first step towards sample preparation. A 3D carbon electrode DEP chip is used to trap and transfer selected cells to the process plate. The final objective is to integrate cell separation and lysis on the automated chip to obtain the intra cellular material for sample processing.

A 3D carbon electrode chip, fabricated by SU8-Photolithography, as detailed by our group previously, [1] is used for this experiment. The chip is mounted on a 3D printer facing the printer platform. The 3d printer is programmed using G-code to control the motion of the chip. The setup is shown in Figure 1a. The chip is first immersed in the cell culture plate, which contains a mixture of cells and particles. A mixture of 10um latex particles and *Candida albicans* is used for this process. At 200 kHz, and 20Vpp, *Candida albicans* are trapped by positive dielectrophoresis whereas the latex particles are not trapped. The chip with the trapped cells is raised from the culture by the 3D printer and is transferred to the wash plate. As, the chip is raised out of the culture plate, the media wets the chip. Some cells and 10um particles remain suspended in the media and get carried along with the chip. In the wash plate, the untrapped cells and the latex particles, acquired due to wetting of the chip are removed. Since the electric field remains 'ON' during this step, the *candida albicans* remain trapped. In the next step, the chip is immersed in the sample plate, where the field is switched 'OFF'. In the absence of the DEP force, the candida cells are released in the sample plate. The process is shown in details in the schematic in Figure 1b. The cells retrieved in the sample plate with DEP and without DEP were counted to determine the concentration of the cells obtained in sample plate. When DEP field is off, the cells get transferred as the water wets the electrode surface. The graph in Figure 1c, shows the difference in the concentration of cells obtained in the sample plate in both cases. The higher concentration in the case of DEP proves that DEP can be used to effectively transfer selected cells.

Current work is to develop the process to completely eliminate the cell transfer due to wetting. The number of cells transferred are being characterized as a function of applied voltage and the concentration of cell culture. The final objective is to develop the device for sample preparation, where with the increase of voltage in the chip and thus the resulting electric field, the cells can be lysed. The intra cellular material after cell lysis can be deposited at the sample

site directly. Use of 3D printer can enable control of the chip motion and hence the cells or intra cellular material can be dropped at specific locations automatically.

Figure 1a) The 3d printer set up with the DEP chip mounted on the holder is shown in the figure. b)The schematic of the cell transfer device shows steps involved in picking and transferring the cells from culture to sample plate c) The plot of log of cell concentration obtained after each cycle is shown in this graph. Number of cells obtained in every cycle with DEP are more than the number of cells obtained in the cycles without DEP. This indicates that the DEP chip can be used to transfer

References:

[1] Martinez-Duarte R., Renaud P., and Madou M., 2011, "A novel approach to dielectrophoresis using carbon electrodes," *Electrophoresis*, **32**(17), pp. 2385–92

501600 Improving the Understanding of Early Stage Amyloid Aggregation Using Microchannel Electrophoresis

Abstract:

Amyloid proteins are involved in a variety of diseases including Alzheimerâ€™s Disease (amyloid beta protein) and Type 2 Diabetes (amylin protein). Several technical challenges exist for the quantitative analysis of different sizes of amyloid proteins present during the early stages of aggregation and thus new technologies are required for enhanced protein separation. Microchannel electrophoretic techniques have emerged as powerful tools for distinction between and analysis of different protein species present at specific points in the aggregation process while photo-induced crosslinking of modified peptides (PICUP) has been used to capture the early, transient oligomeric species. A proteinâ€™s characteristics, including its tendency to aggregate, are determined largely by its primary sequence, and thus variations in an amyloidâ€™s primary structure can be correlated to changes in its aggregation process. Here we show the results of UV-CE analysis of native amyloid beta compared to several variant forms. During native amyloid beta aggregation, three species (monomer, oligomer, and fibril) were observed. Comparing variants to native amyloid beta showed that variants possessing increased flexibility (glycine substituted for leucine) exhibited more rapid change between a larger number of species, while variants possessing decreased flexibility (alanine substituted for glycine) exhibited slower change to the same number of species.

502459 Detection of Activated Src in Human Tumor Samples Using 2D SDS Polyacrylamide Gel Electrophoresis

Abstract:

Dasatinib, an inhibitor of Src family kinases, has been observed to produce marked tumor response in one patient and prolonged stable disease in four others in a small clinical trial involving 34 advanced NSCLC patients. (*1*) The side effects of this drug are harsh. A biomarker showing which patients are likely to respond to dasatinib would be useful. 2D SDS PAGE in combination with phosphotyrosine (pTyr) and Src western blotting was used to detect pTyr Src in a set of 14 human NSCLC tumor samples purchased from a biobank and prepared by homogenization with sodium dodecyl sulfate buffer. A total of 4 /14 tumors showed a pTyr-Src signal indicating active tyrosine kinase. The method of 1D and 2D SDS PAGE has potential to be a biomarker test for active Src.

1. Johnson, F. M., Bekele, B. N., Feng, L., Wistuba, I., Tang, X. M., Tran, H. T., Erasmus, J. J., Hwang, L. L., Takebe, N., Blumenschein, G. R., Lippman, S. M., and Stewart, D. J. (2010) Phase II study of dasatinib in patients with advanced non-small-cell lung cancer, *J Clin Oncol 28*, 4609-4615.

495697 Induced Recycle Flow in a Microchannel Using Electroosmosis

Abstract:

Lab on chip devices involves flow through the channel whose characteristic dimension is less than a millimetre. Recycle flows can be used to increase the residence time of a fluid and improve the mixing efficiency etc. In this work, we propose a novel method to recycle the fluid in microfluidic devices through a combination of electroosmosis and pressure driven flow in an electrolyte solution. The system consists of flow through a main channel and a side channel which is parallel to the main channel. The flow through entire microchannel is pressure driven and an electric field is applied across the side channel to reverse the fluid flow. When the applied electric field is sufficiently large i.e. greater than a critical value, the flow in the side channel is reversed and recycle is induced. The critical value of applied electric field beyond which recycle is induced is investigated as a function of zeta potential. The dependence of pressure and flowrates in the main and channels on applied electric is studied. The effect of geometric parameters such as radius and position of the side channel is analysed on flow behaviour in this work. The simple model proposed here provides valuable insight into the functioning of the device.

499479 Colloidal Nanomaterials-Encapsulated Microcapsules for Biomolecular Sensing

Abstract:

Implantable sensors that detect biomarkers in vivo are critical for early disease diagnostics. While many colloidal nanomaterials have been developed into optical sensors to detect biomolecules in vitro, their application in vivo as implantable sensors is hindered by potential migration or clearance from the implantation site. One potential solution is incorporating colloidal nanosensors in hydrogel scaffold prior to implantation. However, direct contact between the nanosensors and hydrogel matrix has the potential to disrupt sensor performance. Here we develop a hollow microcapsule-based sensing platform that protects colloidal nanosensors from direct contact with hydrogel matrix. Using microfluidics, colloidal nanosensors were encapsulated in polyethylene glycol microcapsules with liquid cores. The microcapsules selectively trap the nanosensors within the core while allowing free diffusion of smaller molecules such as glucose and heparin. Glucose-responsive quantum dots or gold nanorods, or heparin-responsive gold nanorods, were each encapsulated. Microcapsules loaded with these sensors showed responsive optical signals in the presence of target biomolecules - glucose or heparin. Furthermore, these microcapsules can be immobilized into biocompatible hydrogel as implantable devices for biomolecular sensing. This technique offer new opportunities to extend the utility of colloidal nanosensors from solution-based detection to implantable device-based detection

503701 Using Particle-Particle Interactions to Enable Challenging DEP Separations

Abstract:

Insulator-based dielectrophoresis (iDEP) is a powerful electrokinetic (EK) technique that is commonly used for the focusing, enrichment and isolation of biological and non-biological particles. In iDEP devices, arrays of insulating structures are used to create a non-uniform electric field. When dielectric particles are exposed a non-uniform electric field, a particle dipole moment is induced, generating a DEP force on the particles. This force dominates particle movement in highly diluted solutions, where particle-particle interactions are insignificant. However, the influence of particle mutual interactions increases along with particle concentration, since particles are more closely spaced and their dipoles can interact with each other. This contribution presents a systematic approach for the characterization of the EK and DEP velocities of a set of target particles, when additional innocuous particles are used as fillers. The characterization was performed by using the streak-based velocimetry technique in a tapered channel [1]. The dependence of the target EK/DEP velocities on the filler particle concentration, as well as on the filler particle size, was studied. Experiments were carried out to assess how the volume fraction and size of the filler particles can be exploited for the enrichment of target particles/cells. Experiments with a low concentration of yeast cells where performed to demonstrate the potential of this technique for enriching target cells- an essential step in sample preparation for many bioanalytical assessments.

493305 Characterization of Streaming Dielectrophoresis Towards Rapid Particle Separation

Abstract:

Characterization of streaming dielectrophoresis towards rapid particle separation

Monsur Islam, Rucha Natu and Rodrigo Martinez-Duarte

Mechanical Engineering Department, Clemson University, Clemson, SC, USA

Here we present the initial results for the characterization of streaming dielectrophoresis (DEP) towards rapid particle separation. DEP is termed as the motion of a particle while exposed to a non-uniform electric field. In a traditional flow-through DEP experiment, the DEP force (F_{DEP}) dominates over the hydrodynamic drag force (F_{DRAG}) and concentrates the targeted particles on the electrodes. In the case of streaming DEP, the F_{DRAG} is manipulated in such a way that the F_{DRAG} overcomes the F_{DEP} and carries away the DEP-concentrated particles from the electrodes. Hence, concentrated and focused streams of particles are obtained at the end of the electrode arrays (Figure 1a). Here we attempt to characterize and understand how the quality of the streams (i.e. stream width, particle concentration, stream position) depends on various experimental parameters such as electric field, flow rate and electrode configuration. To date, we have performed experiments to characterize the effect of applied voltage and frequency, and flow rate on the stream width of 1 µm latex beads using a carbon electrode DEP device. A potential application of streaming DEP may be rapid cell concentration and separation for clinical diagnostics devices.

The fabrication of the carbon electrode DEP device was detailed elsewhere by our group [1]. For the DEP experiments, a particle suspension of 1 µm latex beads was prepared in distilled water to get a particle concentration of 2.5×10^7 particles/ml. The carbon electrodes were stimulated with a sinusoidal signal. We varied the peak-to-peak magnitude of the signal between 7.5 V and 20 V. The frequency was varied between 5 kHz to 2.5 MHz to target positive DEP of the particles. Flow rates ranging from 25 µl/min to 300 µl/min were used for the experiments. We captured the image of the flow of the particles at the end of the electrode arrays 1 minute after turning on the electric field. The images were analysed to find the stream width using the image-processing software ImageJ. At least three experiments were performed for each combination of the experimental parameters.

Results of the experiments are shown in Figure 1. Three regions are defined in the figures: Trap, Stream and Escape. The "Trap" region indicates the phenomenon when all the particles were trapped on the electrodes and no streams were observed at the end of the electrodes due to much stronger F_{DEP} over the F_{DRAG}. "Stream" represents the conditions when focused streams of particles could be observed at the end of the electrodes due to weak dominance of the F_{DRAG} over the F_{DEP}. "Escape" indicates the incidence when no focusing was observed due to the strong dominance of F_{DRAG} over the F_{DEP}. In Figure 1b, 1c and 1d, it should be noted that in each case the stream gets wider while moving

from "Trap" region to "Escape" region. The F_{DRAG} becomes stronger over the F_{DEP} as we move from "Trap" region to "Escape" region. The stronger F_{DRAG} restricts the focusing of the particles by the F_{DEP}, which makes the streams wider.

The ongoing work is to build a computational model with respect to F_{DEP} and F_{DRAG}. The objective of this model is to predict the occurrence of the stream and its quality by plugging in any of the experimental parameters to the model. A COMSOL simulation is also being used to investigate the effect of different electrode configurations on the streaming DEP. The future work includes design and fabrication of certain chambers downstream of the microchannel to retrieve the focused particles for further processes.

References:

ADDIN Mendeley Bibliography CSL_BIBLIOGRAPHY [1] R. Martinez-Duarte, P. Renaud, and M. J. Madou, "A novel approach to dielectrophoresis using carbon electrodes.," *Electrophoresis*, vol. 32, no. 17, pp. 2385–92, Sep. 2011.

Figure 1: (a) Focusing of 1µm latex beads using positive DEP; Results showing how the stream widths of 1 µm latex beads changes with respect to (b) flow rate, (c) frequency of the applied field, and (d) peak-to-peak voltage.

503698 Effect of Insulator Post Shape on Joule Heating Effects in Insulator-Based Dielectrophoretic Devices

Abstract:

Insulator-based dielectrophoretic (iDEP) devices have shown great potential and versatility for biotechnological applications. Our group has been working on identifying the best design characteristics to increase the efficiency of particle manipulation inside microchannels with insulators. We learned in 2014 [1] that geometries with sharp edges produce an overall higher () that leads to lower voltage requirements to achieve dielectrophoretic trapping. In 2015 we reported a follow up study [2] where we analyzed more in depth the effect of the postsâ€™ geometry on particle trapping by means of dielectrophoresis; this study allowed identifying optimal geometries that produce good particle trapping at much lower voltages. However, as the amount of trapping sites was significantly increased in the optimized geometries (in comparison to conventional geometries), there is a concern as to whether this will also produce an important rise in the sample temperature. This study compares the Joule heating generated in optimized, sub-optimal and standard insulating post shapes by analyzing values of the gradient of the electric field squared, temperature, and magnitude of particle trapping. A 3D model was evaluated, which allowed dissipation of heat for all sides of the microchannel, and experimental validation of the model was performed by employing Rhodamine B, a temperature dependent dye.

1. LaLonde, A., et al., *Effect of insulating posts geometry on particle manipulation in insulator based dielectrophoretic devices.* Journal of Chromatography A, 2014. **1344**(0): p. 99-108.

2. Saucedo-Espinosa, M.A. and B.H. Lapizco-Encinas, *Design of insulator-based dielectrophoretic devices: Effect of insulator posts characteristics.* Journal of Chromatography A, 2015. **1422**: p. 325-333.

502660 Shear-Enhanced Microfluidic Platform for Antibody Purification, in-Situ Efficacy Testing, and Bio-Diagnostics

Abstract:

Selectivity is often the dominant term that is the cause of worry for diagnostic devices for either point-of-screening or point-of-care devices. It is worthwhile to note that similar worries are manifested for the purification of antibodies from cell culture fluid. To separate, filter and purify antibodies the selectivity of Protein A is altered by using pH gradients or different ionic solutions. However, this often leads to post modifications in the purified antibodies leading to reduced efficiency. This becomes paramount for monoclonal antibodies (mAbs) whose extraordinary binding affinity and specificity are used to treat multiple diseases from cancer to cardiovascular diseases. Hence, there is an urgent need to develop processed that allow for antibody purification without using a harsh purification protocol. Here, we show an affinity based lab-on-a-chip microfluidic platform to purify mAbs without post-modification damages, unlike current processes. Further, the inbuilt Electrical Impedance Spectroscopy on this chip, allows for a scrutinization of the binding efficiency of the mAbs antigen pair, replacing diffusion-limited, operator dependent and costly antigen microarray assay or enzyme-linked immunosorbent assay (ELISA). The lab-on-a-chip device has interdigitated electrodes with a channel packed with nanotubes and nanofibers that form a dense matrix. The high packing density allows for modulation of the shear stress in the device. At low flow rates, the mAbs are captured from the cell culture by the matrix post cell lysis, while at high velocities the attached antibody is stripped away by the shear force. Finite elements simulations of the fluid characteristics in the packed device show shear forces in the microchannel of the order of the hydrogen bond strength. Multiple materials are being considered for this shear enhanced platform from Carbon Nanotubes, Carbon Nanofibers to Graphene Oxide. We believe that the control over the shear force will strengthen if a 2D material is taken. This binding and stripping event of the mAbs to the CNT-antigen complex triggers changes in electrochemical impedance spectroscopy (EIS) signal observed from the interdigitated electrodes. The EIS signal is interpreted to obtain a measure of the binding strength and structural changes of the mAbs. Further, the aliquot of purified antibodies is tested using Western Blot and ELISA to show that they are as and often more viable than those collected using other techniques. The tuning of the shear force by the flow allows us to also use this platform as a diagnostic device. Further, the dominant convective transport in the pores, the high sensitivity from the packed matrix, and interdigitated micro electrodes ensures a high signal to noise ratio. This allows for sensitive detection of different biologicals from target DNA to a sandwitched ELISA. As a diagnostic device, we are developing a point-of-screening device that makes it possible to test for cancer panels like Ovarian Cancer. This platform also detects pharmaceutical molecules, like Ibuprofen, Methotrexates, and other NSAIDs in water. Different nanomaterials will be chosen so that the users can use spectroscopy techniques like Infrared and Raman to detect any structural changes of the monoclonal antibodies. The modified platform will allow us to observe the effect of force on the mAbs-antigen binding and resultant physio-chemical structural changes.

492413 AC Droplet Digital PCR

Abstract:

We have developed an AC electrospray droplet generation technology that generates tunable mono-dispersed micron-size water droplets in an immiscible oil at a much higher throughput than any existing droplet generation techniques. The technology also allows easy tuning of droplet size and generation rate to permit large dynamic range quantification of nucleic acids. It enables a novel droplet-digital PCR platform, with high sensitivity, high dynamic range and absolute quantification capability without resorting to Poisson statistics. The droplet generation unit consists of a bundle of micropipettes with conic tips. The sample and PCR mix is pumped through each micropipette and are pinched into droplets at the tips by AC Maxwell pressure without jet formation or Rayleigh instability. A co-flowing or cross-flowing oil phase then carries away the water droplets. To focus the electric field to produce sufficient Maxwell pressure to overcome the large capillary force of small droplets, conic micropipettes are pulled from glass capillaries with a tip diameter of about 10 Î¼m[1]. The frequency is tuned so that Taylor cones and tip-streaming droplets do not appear. The oil phase contains surfactants that stabilizes the emulsion but also allows droplets to pinch off from the micropipette. The droplet size can be easily tuned from 1pl to 1nl by adjusting the electric field and pressure.

This AC droplet digital PCR performs absolute quantification by distributing the templates into a far larger numbers of water droplets, such that each droplet will have only one or zero DNA templates. This binary distribution provides significantly more accuracy than the traditional Poisson distribution of templates in current ddPCR technologies, which typically yield multiple templates per droplet. The presence of a template in each droplet is registered by intercalating dye or Taqman optical reporters after PCR amplification. The thermal cycling is performed either in a batch format with a commercial thermal cycler or with an on-chip PCR unit. To achieve the goal of processing millions of droplets, an automated imaging and counting system with a moving stage and a control software is used. The software coordinates the camera action and stage movement to cover the whole area of the imaging chip. To facilitate the imaging, droplets are densely packed into a single chamber with on-chip pillars that do not allow droplet passage. Validation of this new technology is achieved by bench-marking against traditional qPCR with identical samples.

[1] Pan, Zehao, et al. "Universal Scaling of Robust Thermal Hot Spot and Ionic Current Enhancement by Focused Ohmic Heating in a Conic Nanopore." Physical Review Letters 117.13 (2016): 134301.

504350 Translating Prototype Research from Lab to Commercial Product

Abstract:

Advances in research related to single molecule detection, biomarker discovery, and bio-separations using electrokinetics are promising a grand future for rapid point of care diagnostics. Utilization of customized microfluidics and nanoscale geometric features are enabling unprecedented techniques for sample processing. These advances are great for proof-of-concept hypothesis testing in academia, but translating these devices to the clinic is a major challenge. Designing for economical large scale manufacturing is not considered in the research lab. This critical step in the productization and commercialization process has left many great laboratory based advances not translatable to the clinic. Understanding the requirements, challenges, and techniques available for large-scale manufacturing and the quality control procedures and incorporating them early in the hypothesis testing phase of academic research will guarantee a better chance of productization and translation to the clinic.

506878 Point-of-Care Detection of Hematocrit in a Microfluidic System Integrating Microfabricated Carbon Electrodes

Abstract:

This work describes an electrochemical microfluidic system for point-of-care detection of hematocrit. Hematocrit, the volume percentage of red blood cells in whole blood, is an important medical indicator. Its deviations from normal ranges (40â€“54% for men and 36â€“48% for women) are symptoms of various medical conditions such as anemia and leukemia.

We fabricated carbon microelectrodes via a microfabrication strategy. Carbon precursors, negative photoresist SU-8, were photolithographically patterned on quartz substrates and pyrolyzed at 900 ^{O}C. Specific pyrolysis conditions were optimized to obtain carbon electrodes from large SU-8 features (greater than 10 mm in length) containing more internal stress and residual gases. The SU-8 features were hard-baked at 200 ^{O}C under vacuum pumping to completely evaporate residual gases and to enhance the adhesion. Then, the hard-baked SU-8 were pyrolyzed via a conventional two-step pyrolysis: 1) hard-baking at 200 ^{O}C for 30 mins and 2) pyrolysis at 900 ^{O}C for 1 hour. SU-8 pyrolyzed carbon electrodes were combined with a single PDMS microchannel to complete a device.

To test the feasibility for detection of hematocrit, current responses of red blood cells suspended in phosphate buffered saline were measured at 100 V for 30 secs. A linear trend line of current versus red blood cell concentration was obtained. The results indicate that our device can determine hematocrit within 3.67 % precision and 3.97 % accuracy. Excellent performance of carbon electrodes was attributed to their surface properties. Compared to metal electrodes, carbon electrodes have less bio-fouling tendencies. Large electrode surface areas with porous structures also increase the system voltage efficiency by decreasing the charge transfer overpotential at electrode/electrolyte interfaces.

503582 Nanoplasmonic Biosensors: From Innovative Materials to Multimode Sensing with Integrated Microdevices

Abstract:

Gold nanostructures are a highly attractive class of materials with unique electrochemical and optical sensing properties. Recent developments have greatly improved the sensitivity of optical sensors based on metal nanostructured arrays. We introduce the localized surface plasmon resonance (LSPR) sensors and describe how its exquisite sensitivity to size, shape and environment can be harnessed to detect molecular binding events. We then describe recent progress in three areas representing the most significant challenges: integration of LSPR with complementary electrochemical techniques, long term live-cell biosensing and practical development of sensors and instrumentation for routine use and high-throughput detection. As an example we will demonstrate a novel refractive index and charge sensitive device integrated with nanoplasmonic islands to develop nano-metal-insulator-semiconductor (nMIS) junctions. The developed sensor facilitates simultaneous detection of charge and mass changes on the nanoislands due to biomolecule binding. A brief insight on microcontact printing to functionalize proteins on nanoplasmonic sensors will also be discussed. The developed nanosensors can readily be adopted for multiplexed and high throughput label-free immunoassay systems, further driving innovations in biomedical and healthcare research.

504602 Novel Consumables for High-Throughput Screening That Capitalize on Electrical Forces and Leverage Existing Laboratory Tools

Abstract:

Novel consumables for high-throughput screening that capitalize on electrical forces and leverage existing laboratory tools

Vincent T. Remcho

Department of Chemistry, Oregon State University, Corvallis, OR 97331 USA

Abstract

Plate readers and electrophoresis are to biology and biochemistry what gas chromatographs are to petrochemistry: reliable, indispensable high-throughput sources of critical data. Plate readers have permeated the high throughput screening market just as inkjet printers permeated the printing market; both are high precision tools that handle small volumes of fluids, both have modest acquisition cost, and both are heavily dependent on consumables. Likewise, the quality of their output is dependent on the quality of the consumables.

We have explored a space that is common to plate readers, inkjet printers and bioassays by designing, building and deploying microfluidic paper-based analytical devices (mPADS) using a materials palette that leverages the best qualities and capabilities of each of the materials and methods selected. The *materials* include microfibrous media, polycaprolactone, and semiconductor nanocrystals, and the *methods* include electrospinning, thermal inkjet deposition, lamination, paper electrophoresis, and plate reader-based optical spectroscopy. These selections enable high-throughput screening and analysis with minimal reagent handling and at low cost. Minimization of reagent handling leads to increased precision and accuracy, contributing to data quality and diagnostic utility.

Hybrid microfluidic devices with open and wicking channels were modeled, designed, prepared and applied for colorimetric and fluorescent diagnostic assays for clinical biomarkers such as glucose, bilirubin, and creatinine as well as targets of forensic analysis such as cyanide and heavy metal ions, in environmental and biological samples. Assay results were analyzed using commercial UV-Vis and fluorescence plate readers for detection using custom-built devices of a form factor readily accommodated in unmodified standard plate readers.

504907 Analysis of Single Nucleic Acid Molecules in Micro- and Nano-Fluidics

Abstract:

Nucleic acid analysis has enhanced our understanding of biological processes and disease progression that helps elucidate the association of genetic variants and disease. The need for rapid and high sensitivity nucleic acid detection continues to grow for applications where providing timely results with limited resources remains challenging. Microfluidic and single molecule methods offer alternative solutions to achieve rapid and high sensitivity nucleic acid detection. Solution-based single molecule detection strategies enable high detection sensitivity without the need for nucleic acid amplification and are instead limited only by the volume of sample that can be efficiently screened. Such detection methods couple nicely to microfluidic analysis because microfeatures can be designed to match the fluidic analysis volumes with the single molecule detection volume for highly efficient sample screening. We have demonstrated that the coupling of microfluidics and cylindrical illumination confocal spectroscopy (CICS) enables highly sensitive detection of genetic and epigenetic makers in liquid biopsy. By further coupling CICS with free solution hydrodynamic separation (SML-FSHS), we can size separate individual nucleic acid fragments independently of fluorescence and provide highly quantitative analysis over a large dynamic range by counting single molecule bursts.

505111 New Paradigms in Gel Electrophoresis with Non-Newtonian Fluids

Abstract:

The evaluation of biomolecular structure and function is critical to successfully implementing biotechnology and to authenticating biological reagents that are fundamental to health-related research. Phospholipid nanogels are promising non-Newtonian separation matrices, which are thermally responsive, with a liquid-to-gel transition at ~22 °C. This property makes it easy to load and pattern nanogels in narrow microfluidic channels or narrow-bore capillaries at low temperature. The fluids are then locked in place by raising the temperature to gel the material. Discrete zones of nangels are maintained in microscale channels to accomplish traditional gel-based separations but in a microscale format. For example, nanogels of different viscosity can be patterned to create stacking zones used in DNA sieving and then thermally erased. Nanogels have also displayed a stabilizing effect to enzymes and other proteins, which provides the means to incorporate different affinity reagents into the separation. This enables cost-effective use of nanoliter volumes of highly specific enzymes and antibodies in microscale channels to identify specific proteins as well as post translational protein modifications. A wide range of applications will be discussed to demonstrate this new approach to translate traditional gel-electrophoresis separations into a microscale format.

503625 Entropic Trapping during DNA Transport in Microfluidic Gel Electrophoresis: A Monte Carlo Simulation

Abstract:

The ability to precisely manipulate transport of charged macromolecules (DNA, proteins) in nanoporous surroundings plays a critical role in the design of microfluidic electrophoresis systems. Reptation is the predominant transport mechanism in most hydrogel-based separation matrices, where the characteristic macromolecular size scale is larger than that of the surrounding pores. DNA electrophoresis in the reptation regime yields a scaling of mobility with the inverse of macromolecular size. Alternatively, when the characteristic macromolecular size is the same as that of the surrounding pores, transport occurs via an entropic trapping mechanism. In contrast to reptation, transport in this regime occurs via discrete hops between larger pores joined by narrow interconnecting spaces, leading to a stronger size dependence of electrophoretic mobility that is desirable for enhanced separation performance.

We have previously shown how application of a time varying electric field driving force can enable the activated hopping process to become synchronized, leading to a resonance state that can be exploited both to obtain improved separation performance and to extract nanoscale physical parameters of the macromolecules (contour length, persistence length) [1, 2]. In these initial studies, our entropic trapping transport model was focused on the activated process by which molecules enter the narrow space between adjacent large pores (activation time), but assumed a simplified representation of transit through the confined interconnecting region (migration time). Specifically, migration was described in terms of a simple function of the macromolecule's radius of gyration and the applied electric field. Although this representation is consistent with previous frameworks that have been developed to describe electrophoretic entropic trapping, it does not realistically capture the hydrodynamic forces and conformational changes that occur during migration through the confined interconnecting pore spaces.

Here we apply a more realistic representation of migration through the nano-confined interconnecting spaces linking adjacent large pores. Our starting point is a model is based on a Monte Carlo algorithm developed by Slater, et. al. [3] that directly incorporates frictional and entropic forces. The portion of the macromolecule located outside of the pore space experiences frictional interactions with the surrounding fluid, while the portion inside the pore experiences drag due to spatial confinement in the narrow interconnecting space. Entropic forces are related to conformation differences between the portions of the macromolecule at the entrance and exit of the pore space. Embedding each of these effects into our model enables the size dependence of electrophoretic mobility to be computed by integration over the pore size distribution of the hydrogel matrix. The resulting predictions are validated by comparison with experimental data. These new insights can suggest optimal operating conditions (electric field amplitude and frequency) and nanoporous gel matrix architectures (pore size distribution) that deliver enhanced separations in microfluidic electrophoresis systems.

[1] Shi, N. and Ugaz, V.M. "Noise Enhanced Gel Electrophoresis." *Electrophoresis*, 35 (2014): 1758-1765.

[2] Shi, N. and Ugaz, V.M. "An Entropic Force Microscope Enables Nano-scale Conformational Probing of Biomolecules." *Small*, 10 (2014): 2553-2557.

[3] Gauthier, M.G. and Slater, G.W. "A Monte Carlo Algorithm to Study Polymer Translocation Through Nanopores. I. Theory and Numerical Approach." *J. Chem Phys.*, 128 (2008): 065103.

503699 Obstacle-Density Effects on Particle Trapping in Insulator-Based Dielectrophoresis Systems

Abstract:

In insulator-based dielectrophoretic (iDEP) devices, insulating structures are placed between electrodes to produce non-uniformities in the electric field. This creates polarization effects on particles and allows for their manipulation when the dielectrophoretic force induced overcomes all other forces in the system. Microchannel designs usually include an array of insulators, *i.e.* posts, and particle trapping sites are generated within the post-array. An increase in the number of posts enlarges the number of potential trapping zones where particles can be immobilized. This study presents the effect that the number of columns of insulating structures have on the gradient of the electric field squared in an iDEP microdevice. A mathematical model was created in COMSOL Multiphysics and later validated by experimentation, where it is shown that the column density (number of columns/unit length) has an effect on the magnitude of the dielectrophoretic force produced when an electric field is generated in a microdevice.

498857 Investigating the Impact of Low Concentration Surfactant on Red Blood Cell Dielectrophoretic Responses

Abstract:

Red blood cell dielectrophoretic responses have been extensively studied for blood typing, sepsis and circulating tumor cell separations. Surfactants are a common pretreatment to prevent sticking within dielectrophoretic microdevices. In this study, the dielectrophoretic motion of untreated human erythrocytes (red blood cells, RBCs) were compared to RBCs treated with low concentrations of Triton X-100, which is a widely used nonionic surfactant to solubilize integral membrane proteins. However, at relatively low concentrations, this surfactant has been shown to effectively stabilize cells against osmolysis. This biphasic behavior has not been fully understood, especially within cells subjected to AC electric fields. The objective of this work is to characterize the interaction between Triton X-100 at low concentrations (0.07-0.22mM) and the red blood cell membrane by quantifying changes in the cellâ€™s dielectrophoretic crossover frequency (f_{CO}). The AC frequency at which cells change the direction of their movement from negative DEP to positive DEP was determined by applying a 5 V_{pp} signal with a frequency sweeping from 700 to 300 kHz in four pairs of converging triangular (planar Cr/Au) electrodes. At relatively low concentrations (0.07mM), the crossover frequency f_{CO} was lower compared to untreated RBCs. The f_{CO} for the control cells averaged 488 kHz, while the treated cells' f_{CO} was 390 kHz, suggesting that at 0.07mM Triton X-100, the relative membrane permittivity was increased. A plausible mechanism for the amphiphilic surfactant is insertion into the bilayer which alters ion permeability. Altered ion permeability would cause asymmetric component distribution between the outer and inner membrane surfaces disrupting the ion concentration equilibrium between the interior and exterior of the cell membrane. As the concentration of the surfactant increased to 0.14mM, cells began to exhibit negative-DEP during the entire experiment, while others exhibited a f_{CO} of 390 kHz. At this concentration of Triton X-100, cell viability was not compromised. Thus, experiments were performed to determine cell membrane integrity after two hours of treatment. When the concentration of the Triton X-100 reached the critical micelle concentration (0.18 to 0.22 mM), cells uniformly exhibited negative-DEP over the entire frequency range. This work provides mechanistic understanding into the chemical interactions between surfactants and cell membranes that can subsequently be exploited to enable precise control for electrokinetic microdevice separations and operation.

496609 Modeling of Nano-EIS in a High Peclet Number Packed Microfluidic Biosensor

Abstract:

Microfluidic Biosensors are characterized by both, low Peclet and Reynolds number laminar flow. Traditionally the three well-known transduction methodologies used in microfluidic biosensors are electrochemical methods, optical methods including fluorescence and electrical impedance-based techniques. Here we demonstrate the different physics in the electrochemical and electrical impedance based (EIS) transduction method arising from a microfluidic geometry that promotes convective transport. This is achieved by packing Carbon Nanotubes (CNTs) into a microfluidic channel. The high packing density of the CNTâ€™s promotes convective transport. This is verified using cyclic voltammetry (CV), using ferric ferricyanide in 0.1 X PBS buffer. Further, the observed CV curve for our microfluidic system shows a pattern, unlike traditional CV curves which give us interesting perspective into the ionic reactions at the CNT interface. CV curves are plotted for bare CNT, CNT surface activated using NHS-EDC chemistry and with attached monoclonal antibody to CNT surface in 0.1x PBS aqueous at scan rates of 25 to 200 mV/s. It is interesting to note that in the absence of CNT, our device demonstrates regular CV characteristics where the peak current analysis is carried out using the Nicholson method. However, post CNT packing in the device, the CV curve deviates from the traditional CV curves. A thorough analysis of the CV indicates that convective transport dominates over diffusion. Further, the analysis clearly shows the interplay between surface coverage and charge transfer which results in different peaks in the CV sweep from CNT, CNT surface activated using NHS-EDC chemistry and with attached monoclonal antibody to CNT surface. The two competing phenomenon of surface coverage and charge transfer is illustrated in more details using electrochemical impedance spectroscopy or EIS. On addition of a secondary antibody to the monoclonal antibody-CNT system, it was observed that the EIS shift (chiefly the charge transfer resistance) decreased from bare CNT to EDC activated CNT to monoclonal antibodies and increased on the addition of secondary antibodies. Current literature studies show that on the addition of monoclonal antibody and then a polyclonal antibody always leads to an increase in the charge transfer resistance due to a decrease in the electrode surface area available to the buffer. This occurs from the increased electrode surface coverage from the large antibody molecules. It is important to note that, CNTs, as they are touching the electrodes on the glass surface, behave as extended electrodes here. Here, the addition of more charge to the CNT surface from bare CNT to EDC-NHS surface activated CNT to antibody functionalized CNT decreases charge transfer resistance. Initially, the addition of charge dominates (decrease in charge transfer resistance) while on the addition of the secondary antibodies, charge transfer resistance increases due to the decreasing CNT surface area. We ascribe this departure from classical formulations based on the nano-ordered packing structure of the CNT inside the channel. Here, the nanostructured CNT results in the ionic flux being confined in local nanodomains. This is modeled using a new EIS circuit which also allows for a new convective transport path. The EIS model further predicts the change in the EIS measurements due to the attachment of the monoclonal antibodies in the nano-ordered CNT structure. This study paves the way for increasing the sensitivity and selectivity of a packed biosensor.

503700 AC Electrohydrodynamics of Polarized Laminar Flows

Abstract:

In this work, we explore the fluidic dielectrophoresis (fDEP) behavior of laminar flows subjected to an external alternating current (AC) electric field. In fDEP, an AC field is applied across a liquid electrical interface created between two co-flowing streams. We show that the field drives an interfacial displacement that is dependent on the field frequency and that relative differences in electrical conductivity and dielectric constant between the two liquid flows. At low frequency, this deflection is independent of dielectric constant, while at high frequency it is independent of electrical conductivity. At intermediate frequencies, we observe a cross-over frequency that is sensitive to both fluid electrical properties, and where no fluid displacement is observed. We present an electrohydrodynamic polarization model that accurately predicts this electrical dependence. Furthermore, we experimentally validate this model using high-speed imaging and capture the interfacial fluid velocity of the displacing interface. We show that the interface velocity can be well-predicted by a simple uni-directional transport model. Finally, we demonstrate the ability to rapidly measure fluid properties, including liquid dielectric constant and viscosity at low-cost using a simple microfluidic device.

497550 Micro Electrokinetic Turbulence and Its Measurement in a Microchannel

Abstract:

We first report some phenomena on microelectrokinetic turbulence, which is experimentally achieved in a microchannel at relatively low Reynolds number (on the order of 1 without forcing) when a pressure driven flow is electrokinetically forced. The measured maximum Reynolds number under forcing is not higher than 100. The inlet of the flow has two streams with different electrical conductivity, and the ratio of the conductivity is 1:5000. To characterize the flow, measurement of the statistical features of the flow is required. However, conventional velocimeters, such as microPIV that uses particles as tracer, cannot measure such unsteady flows with sufficiently spatial and temporal resolution simultaneously. In addition, in electrokinetic flows, the particles have essentially different velocity from their surrounding flow. Therefore, here we apply laser induced fluorescence photobleaching anemometer (LIFPA), which uses an electrically neutral charged, small molecular dye, to measure the flow velocity to avoid these issues. Home-developed confocal LIFPA enables required simultaneously high spatial and temporal measurement. The observed microelectrokinetic turbulence can generate ultrafast mixing and LIFPA would become an promising velocimeter for accurate measurement of flow velocity in complicated flows in microfluidics, where the velocity of particles are different from that of the fluid, and simultaneously high spatial and temporal resolution is required.

499722 Thermodynamic Modeling of Electrodes and Mobile Ions in Capacitive Deionization Cell Units with Enrtl Model

Abstract:

Capacitive deionization (CDI) is a technology to purify brine water of low concentration with cell units. The two electrodes in classical CDI unit attract opposite charged ions in the brine water when charged and release them when discharged. Using electrodes modified with immobile chemical charge, inverted CDI units take in ions when not charged while release them during charging. Biesheuvel et al. put forward a dynamic model of the electrical double layer and electrodes, however, thermodynamic activity of ions in the electrodes was not considered. [1,2]

We have established a thermodynamic framework for polyelectrolyte solutions by introducing Manning's limiting law to the eNRTL model. For a polyion that has a charge density larger than certain critical value, the counterions in the polyion solution will condense on the surface of the polyion to reduce the charge density. The degree of the condensation is positively related to the charge density of the polyion. Manning first described the counterion condensation phenomenon by integration of the potential between mobile ions and polyions as the limiting law. [3] The eNRTL short-range equations are used to describe the interactions between particles in their immediate neighborhood. [4] Our model is a comprehensive one that works not only in the limiting case but also any multi-component system containing polyelectrolytes with high accuracy, including ion exchange membranes and resins.

The classical and inverted CDI units at equilibrium have the electrodes with counterions electrically or chemically attached to the surfaces. The scenario is very similar to the counterion condensation process of polyions. We are applying our thermodynamic framework to calculate the equilibrium adsorption capacity of CDI systems. Consider the electrodes and attached ions as polyions. The polyions form separate phases from the bulk simple salt solution. These polyions phases can be treated as the polyelectrolyte-water binary cases.

By equality of chemical potential of solute (NaCl) in the bulk solution and the electrode phase, we have

$$m_{Na^{-bulk}}\gamma_{Na^{+bulk}}m_{Cl^{-bulk}}\gamma_{Cl^{+bulk}} = m_{Na^{-electrode}}\gamma_{Na^{+electrode}}m_{Cl^{-electrode}}\gamma_{Cl^{+electrode}}$$

where m is the molality of species; γ is the activity coefficient.

With this equation, we are able to relate concentrations of bulk solution, amount of adsorbed salt and charge density of the electrodes. The activity coefficients in the bulk solution can be calculated with the original eNRTL model while the ones in the electrode phase need to be computed with the eNRTL model modified with Manning's limiting law to account for the "point-to-line" electrostatic potential. In the electrode phase, the two types of ions are both considered as counterions since oppositely charged ions are attracted at the two electrodes. Equilibrium salt adsorption and cell voltage data are used to fit the charge density of the electrodes as the parameter in Manning's limiting law. The short-range interaction parameters between water and the charged segments on the electrodes also need to be regressed from data.

Modeling results show that higher charge density on the electrode will improve the performance of CDI units. We are also interested in the relationship between the cell voltage and the polyion charge density value, and finding the working condition of the best performance given certain brine concentration.

Reference

1. Biesheuvel, P. M., et al. "Attractive forces in microporous carbon electrodes for capacitive deionization." *Journal of solid state electrochemistry* 18.5 (2014): 1365-1376.
2. Biesheuvel, P. M., H. V. M. Hamelers, and M. E. Suss. "Theory of water desalination by porous electrodes with immobile chemical charge." *Colloids and Interface Science Communications* 9 (2015): 1-5.
3. Manning, Gerald S. "Limiting laws and counterion condensation in polyelectrolyte solutions I. Colligative properties." The journal of chemical Physics 51.3 (1969): 924-933.
4. Song, Yuhua, and Chau-Chyun Chen. "Symmetric electrolyte nonrandom two-liquid activity coefficient model." Industrial & Engineering Chemistry Research 48.16 (2009): 7788-7797.

504188 3D Dielectrophoresis

Abstract:

Dielectrophoresis of biological entities has been the subject of increasing research density over the last 50 years; publication rates exceeded first 20 papers/year in 1991, increasing to 100/year in 2003, 200/year in 2005 and 400/year in 2009. However, most papers still begin with describing â€œgreat potentialâ€ without this having been yet realised; 2014 and 2016 saw significant drops in DEP paper output, and growth appears to have stabilised. We have been promising great things and the technology now needs to deliver. In order to do this, we need to better understand the requirements that DEP characterisation and separation are needed to fulfil.Â We have developed the 3D dielectrophoresis platform in order to address some of the shortcomings of DEP as perceived by our end-user base; it enables a characterisation platform that is easy to use, and a separator platform capable of separating cells label-free at rates comparable to, or faster than, FACS and MACS. Both use disposables that offer simplicity of manufacture and low user cost. The next step will be to use this to better understand what the DEP response means, and how this can provide the information that electrophysiologists require.

503651 Electrical Manipulation of Cells for Biology and Medicine

Abstract:

Microsystems have the potential to impact biology & medicine by providing new ways to manipulate, separate, and otherwise interrogate cells. Simply physically manipulating cellsâ€"using microfluidics, electric fields, etc.â€"provides new ways to separate cells and organize cell-cell interactions. One example illustrating the power of microscale manipulation of cells is to sort cells based on their intrinsic electrical properties. Electrical properties have previously been correlated with important biological phenotypes (apoptosis, cancer, etc.), but a sensitive and specific method approach has been lacking. We have developed a method called iso-dielectric separation that uses electric fields to drive cells to the point in a conductivity gradient where they become electrically transparent, resulting in a continuous separation method specific to electrical properties. With this method, we are developing a point-of-care assay that can quickly assay immune cell activation, which has applications for monitoring inflammation in sepsis and other immune disorders.

504189 Dielectrophoresis Shows That Regular Potassium Transport Controls the Circadian Electrophysiological Rhythm in Human Red Blood Cells

Abstract:

The standard model of biological rhythmic behaviour (also called "circadian", from the Latin for "around a day") is one based on gene expression cycles; several genes are activated and inactivated in a near-24-hour pattern in nucleated cells, following a well-characterised mechanism. However, this model was challenged when circadian rhythms in PRX activity were observed in isolated mammalian red blood cells (RBCs), which lack nuclei. This suggested the presence of a non-genetic mechanism which operates in RBCs. In order to investigate this mechanism, we employed a combination of approaches to scrutinize the causes and consequences of timekeeping in RBCs. These studies indicated highly rhythmic, sinusoidal rhythms in the membrane conductance and cytoplasmic conductivity of isolated RBCs. Using a combination of approaches including dielectrophoresis and ICP-MS, we have identified rhythms in both K^+-flux and intracellular K^+. We also found that we could speed up, slow down or completely abolish the rhythm in DEP properties by using pharmacological intervention and ion replacement. These results indicate that K^+-transport underlies the circadian rhythms of RBC membrane physiology and suggest that K^+-transport may contribute to the determination of circadian period in isolated RBCs.

503125 Isomotive Dielectrophoresis: Design Considerations and Scaling Laws

Abstract:

Dielectrophoresis (DEP) is a technique in microfluidics that has demonstrated trapping, concentration, manipulating, and sorting of biological cells. However, there has been less emphasis on the development of *analytical* DEP devices. Analytical DEP devices have been rarely pursued due to the inherent non-uniform nature of the dielectrophoretic force, which is proportional to the gradient of the field-squared, thereby generating a highly non-uniform force field within a microfluidic device. Our device overcomes this limitation by applying a unique isomotive dielectrophoresis (isoDEP) field. Decades ago Herbert Pohl first used the phrase â€œisomotive dielectrophoresisâ€ and suggested an electrode geometry for particle sorting applications. We have modified and adapted his solution for a microfluidic isoDEP device where the gradient of the field-squared is constant thereby enabling a constant DEP force to be applied to all particles in the field of view. In combination with particle tracking velocimetry the Clausius-Mossotti factor for each particle can be extracted. With this approach detailed dielectric measurements of each individual cell is possible with a greater throughput than the current practice using electrorotation.

This presentation will describe design considerations and scaling laws associated with two fundamentally different isoDEP platforms. The unique isoDEP analytical solution will be presented. AC electrokinetic products inherent to such systems such as Joule heating, AC electro-osmosis, and electrothermal flow will be discussed. The current status of isoDEP platform automation and proof-of-concept results using biological particles will be presented.

503483 Exploring the '2nd Frontier' of Dielectrophoresis and its Application in the Biomedical Sciences

Abstract:

Dielectrophoresis (DEP) has been widely studied for its potential as a biomarker-free method of sorting cells and other bioparticles based on their intrinsic dielectric properties [e.g., 1-8]. Under appropriate experimental conditions the DEP frequency spectrum typically exhibited by a viable mammalian cell in suspension is characterized by two frequencies, fxo1 and fxo2. At these two characteristic frequencies, commonly known as the DEP cross-over frequencies, the effective electrical impedance (i.e., the a.c. conductance and capacitance values) of the cell exactly match those of the fluid it has displaced.

At low frequencies (<10 kHz) viable cells typically exhibit negative DEP and move away from electrodes, with the transition to positive DEP occurring at fxo1 - where the cells move towards high field regions at the electrodes. The investigation, understanding (in terms of cell biology) and biomedical exploitation of this phenomenon can be considered to represent exploration of the '1st frontier of cellular dielectrophoresis'. The results of this work, extending over more than 25 years, are now exploited in the form of various commercial devices for clinical and biomedical applications. The status of this activity and possible future trends will be reviewed.

A transition from positive DEP back to negative DEP occurs at fxo2 (~150 MHz [10]). The factors that control the high frequency DEP cross-over at fxo2 have only recently been explored, and the status and implications of this '2nd frontier of dielectrophoresis' will also be discussed in this presentation.

1. J. Voldman, Electrical forces for microscale cell manipulation, Annu. Rev. Biomed. Eng. 8, 425 (2006).

2. B. H. Lapizco-Encinas and M. Rito-Palomares, Dielectrophoresis for the manipulation of nanoparticles, Electrophoresis 28, 4521 (2007)

3. R. Pethig, Dielectrophoresis: Status of the theory, technology and applications, Biomicrofluidics 4, 022811 (2010)

4. B. Cetin, D. Li, Dielectrophoresis in microfluidic technology, Electrophoresis 32, 2410 (2011)

5. Z. R Gagnon, Cellular dielectrophoresis: Applications to the characterization, manipulation, separation and patterning of cells, Electrophoresis 32, 2466 (2011)

6. R. Martinez-Duarte, Microfabrication technologies in dielectrophoresis applications: A review, Electrophoresis 33, 3110 (2012)

7. M. P. Hughes, Fifty years of dielectrophoretic cell separation technology, Biomicrofluidics 10, 032801 (2016)

8. R. Pethig, Dielectrophoresis: Theory, Methodology and Biological Applications, John Wiley & Sons, Chichester (2017).

9. C. Chung, M. Waterfall, S. Pells, A. Menachery, S. Smith and R. Pethig, Dielectrophoretic characterization of mammalian cells above 100 MHz, J Electr Bioimp, 2, 64 (2011).

17AIChE
Annual Meeting, Minneapolis, MN

Penn
University of Pennsylvania

Diverse Colloidal Crystals from DNA-grafted Spheres Via Self-assembly

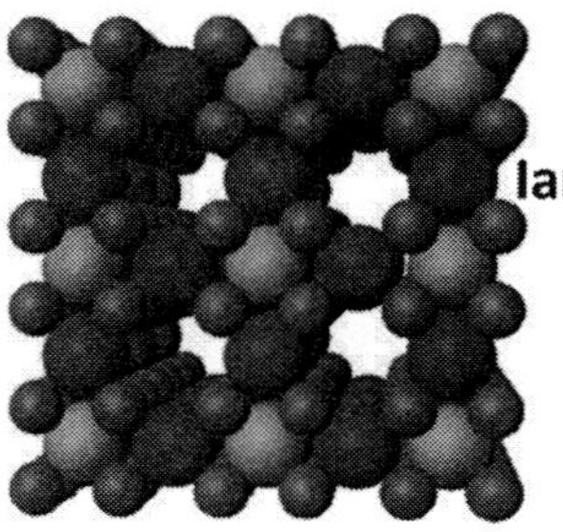

Yifan Wang

Ian Jenkins, Talid Sinno, John Crocker

Dept. of CBE

University of Pennsylvania

Advisor: Prof. John Crocker

AIChE Annual Meeting

November 2nd, 2017

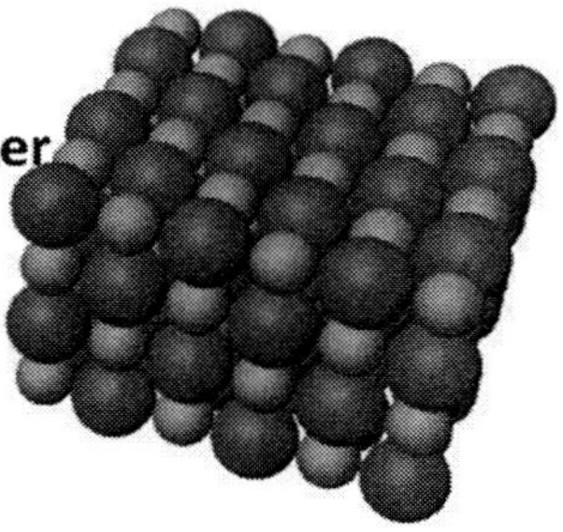

Motivation

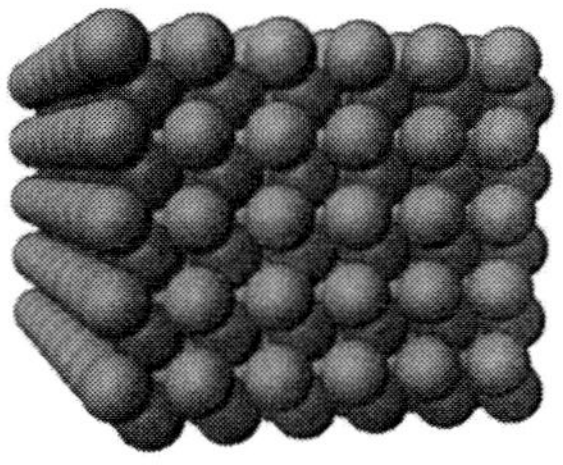

CsCl

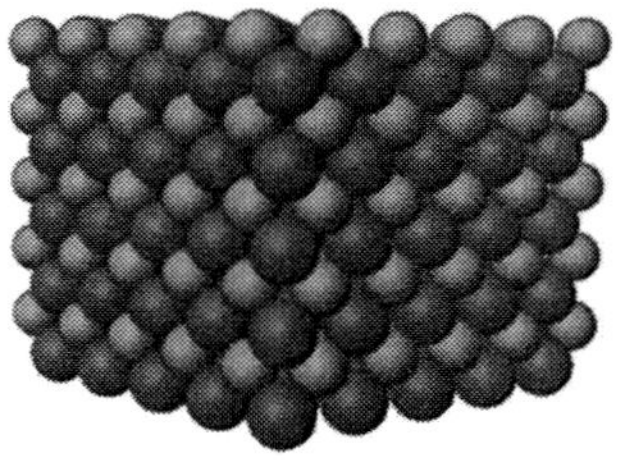

FCC

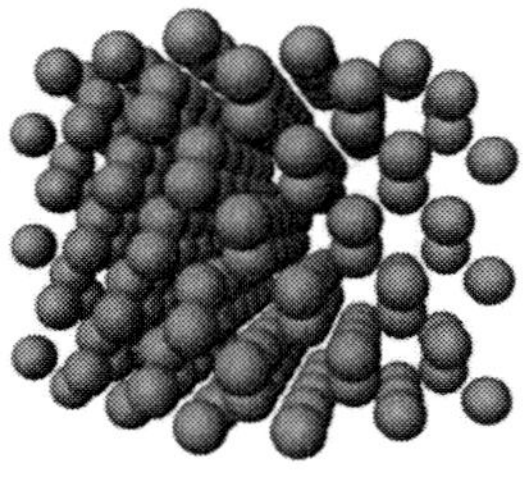

Diamond

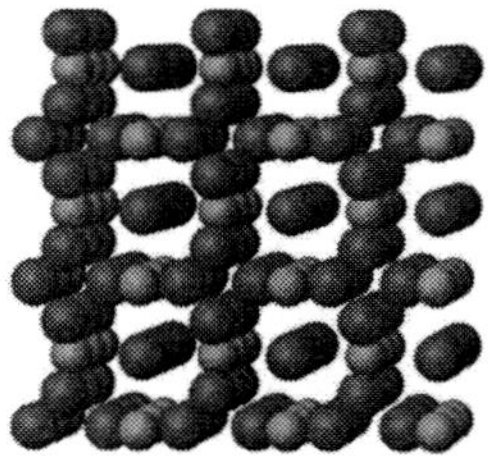

Cr3Si

Can novel crystal structures be made of colloids?	Yes!
How to connect each species particles?	DNA!
Are they stable? What do they transform to?	2 step self-assembly

2

DNA Binding

Rapid Prototyping!!!

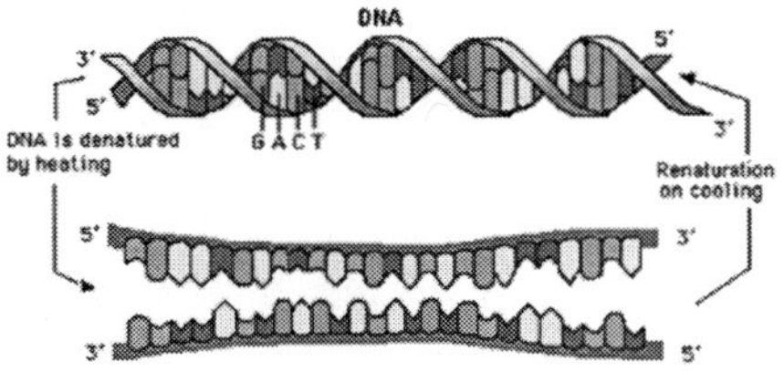

Through DNA linker

TGACGTTCAT
| | | | | | | | | |
ACTGCAAGTA

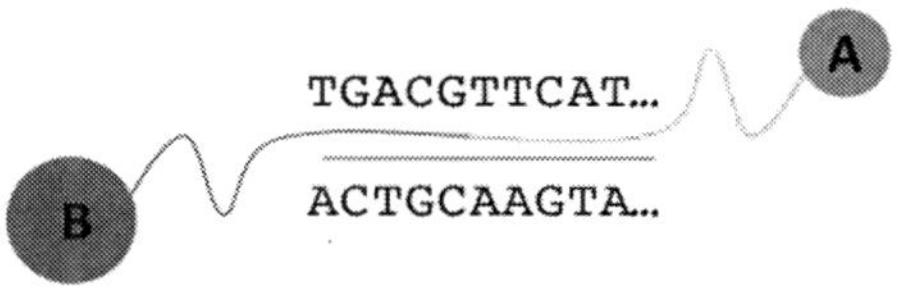

Polystyrene Colloids

Base Pairs Binding

A–T
T–A
C–G
G–C

5′-ATGGCTACGT-3′
| | | | | | | | | |
3′-TACCGATGCA-5′

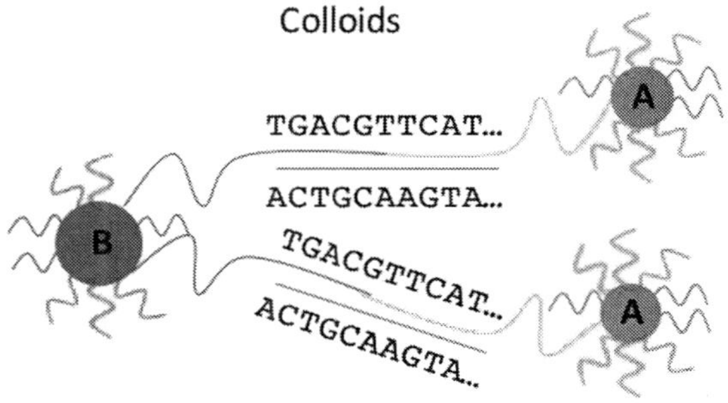

Multiple Particles

3

Chemistry Version I

University of Pennsylvania

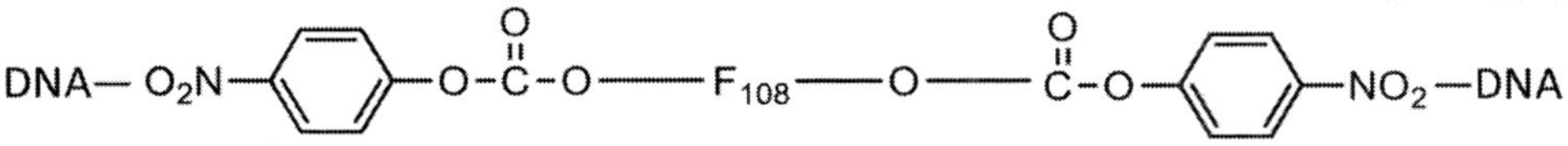

4-NPCF

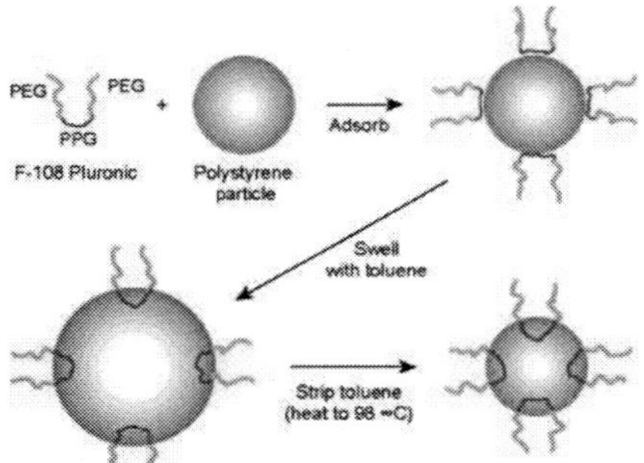

Particle formation is important, particles could **roll and slide** using our method after swell/deswell

Swelling-Based Method for Preparing Stable, Functionalized Polymer Colloids, John Crocker, JACS, 2005

High DNA concentration obtained!!

An Improved chemistry right now!

On the scope, crystals are formed in less than 1 hour

4

2 Step Self-Assembly

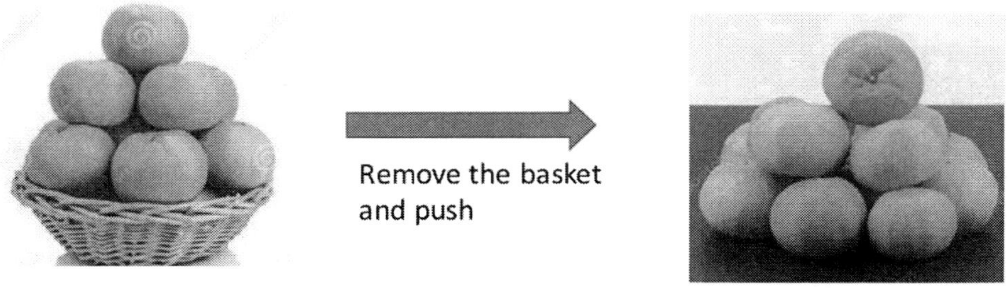

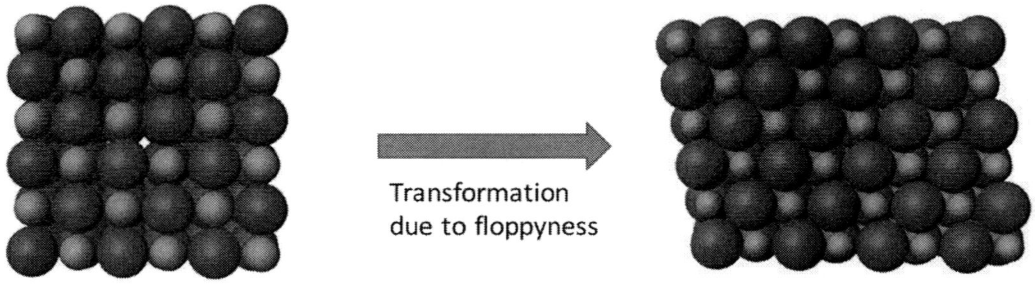

Parent phase
First step

Child phase
Second step

2 step directed particle self-assembly

5

Basic Idea

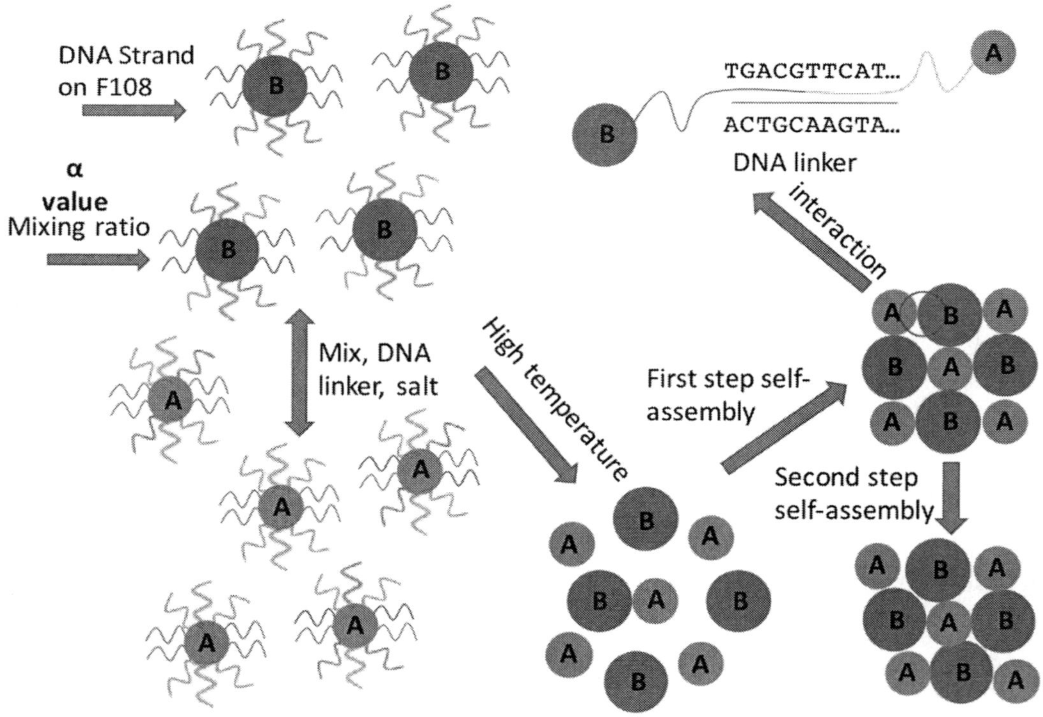

6

Parameters

A	245	378	392	378	378	(nm)
B	445	445	445	392	378	(nm)
Size Ratio:	**0.551**	**0.849**	**0.881**	**0.964**	**1**	

Stoichiometry:

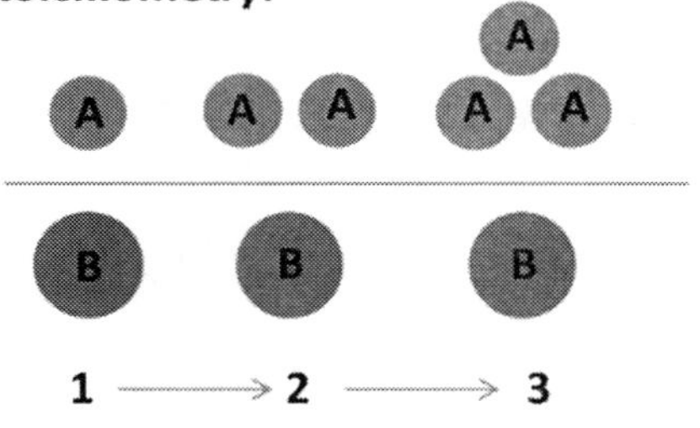

Interaction/Attraction Matrix and mixing ratio

Only A-B	A	B
A		+
B	+	

B-B A-B	A	B
A		+
B	+	+

A-A A-B	A	B
A	+	+
B	+	

A-A B-B A-B	A	B
A	+	+
B	+	+

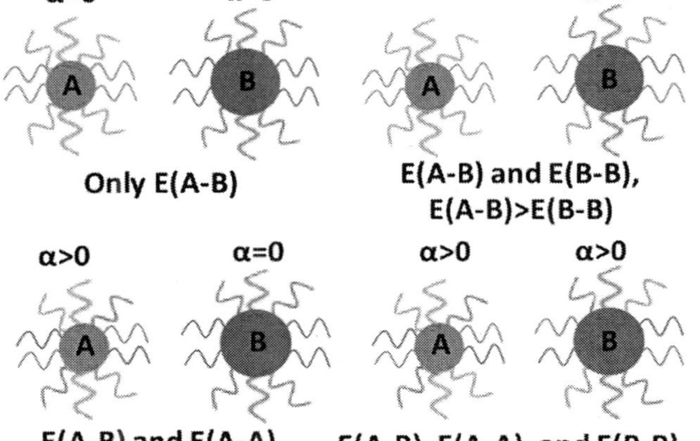

7

Crystallography

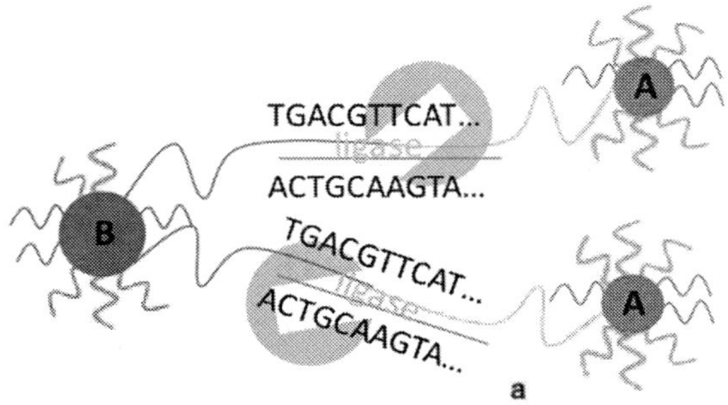

Obtain stable crystal structure

Z-Stack measurement using index match oil, see through the structure, compare different 2-D planes.

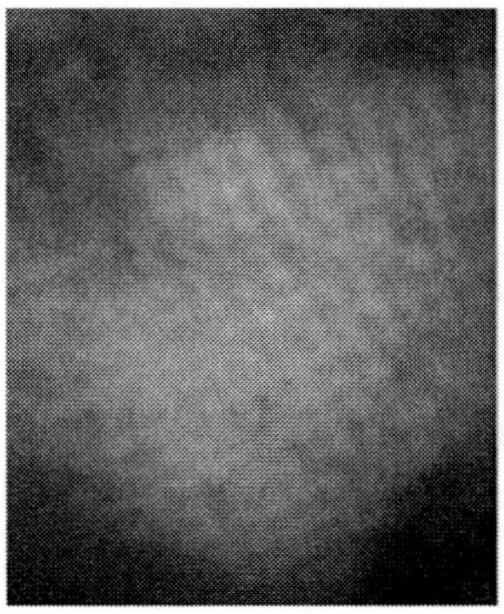

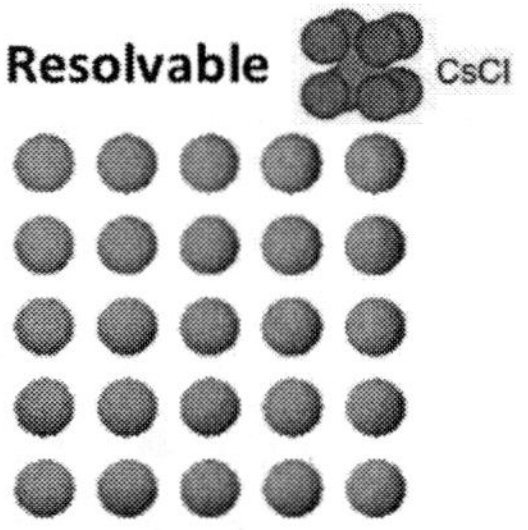

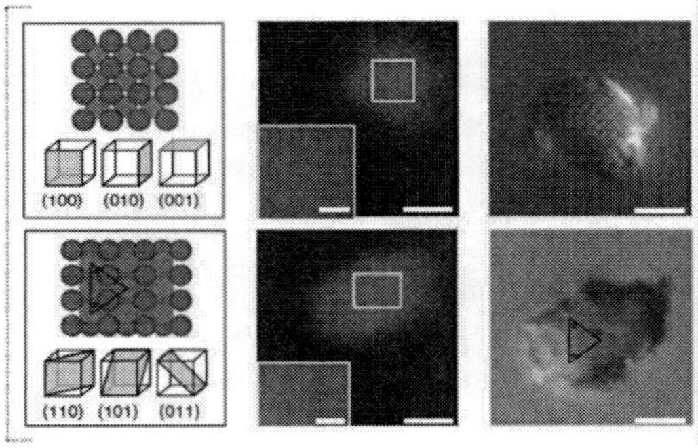

Driving diffusionless transformations in colloidal crystals using DNA handshaking, M. Casey, Sinno, Crocker, et al., Nature Communications, 2012

Experimental Summary

Parent Phase:
a, b, h

Child Phase:
c, d, e, f, g, i, j

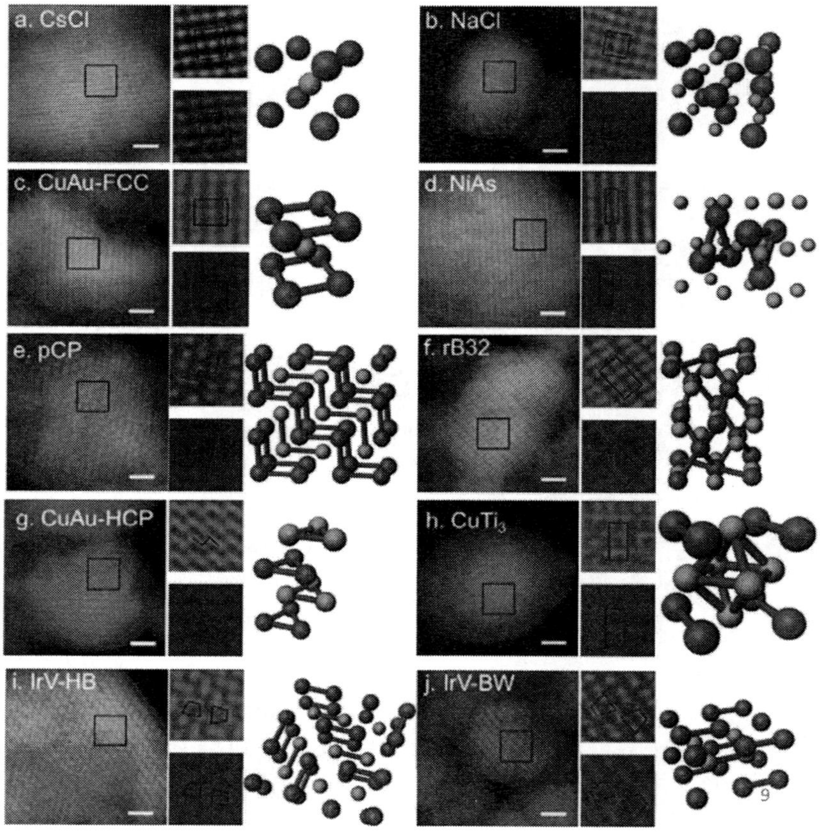

Transformation Paths

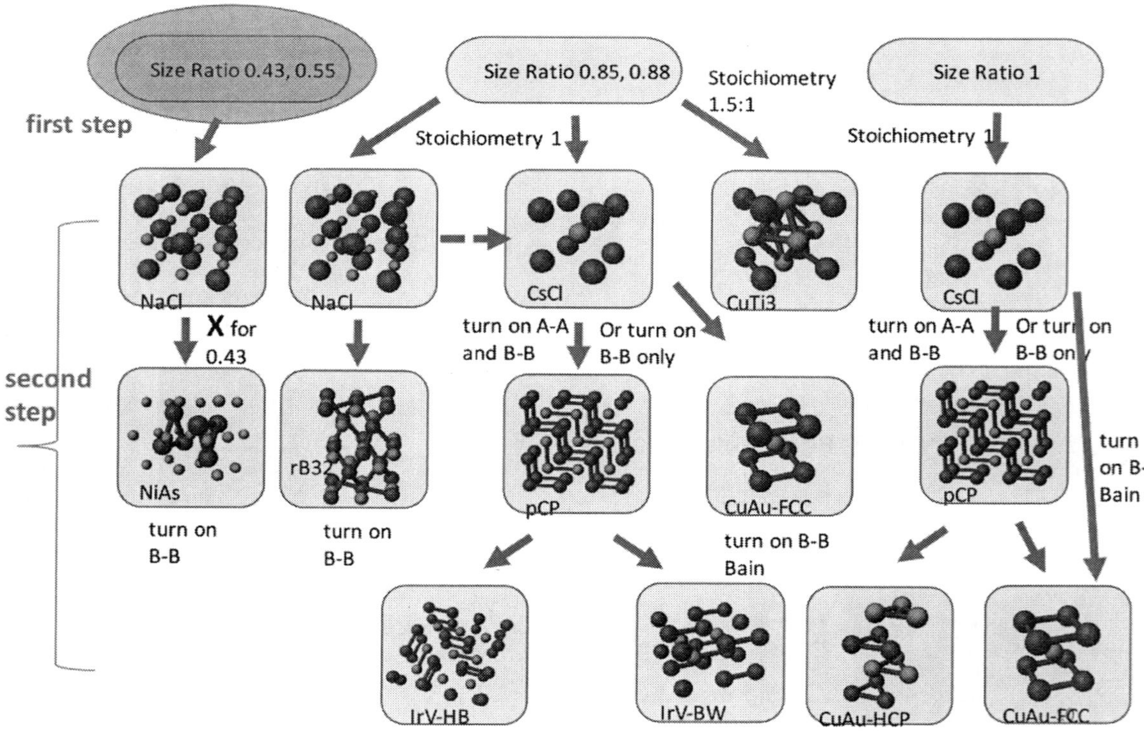

NaCl through first step self-assembly

A 245nm

B 445nm

Size Ratio: 0.551

Stoichiometry: 1

Only A-B	A	B
A		+
B	+	

Perspective view of NaCl

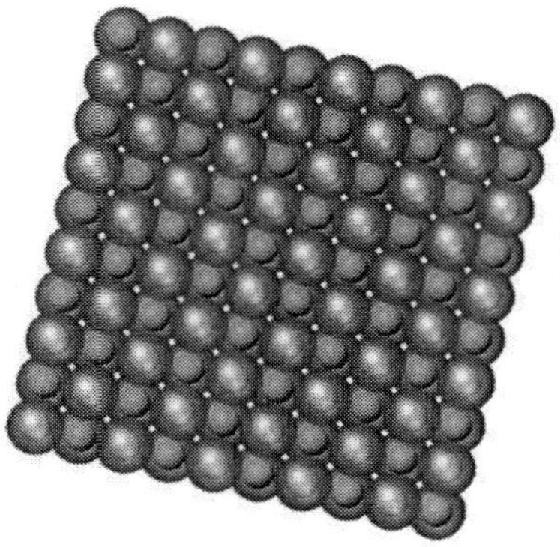

MAYA image of NaCl 100 Plane

Lattice spacing ~380nm

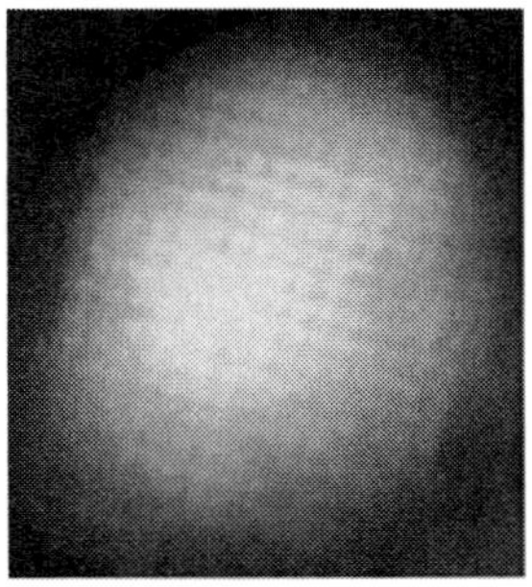

Confocal Image of NaCl 100 Plane

Lattice spacing ~400nm

Trigonal NaCl (NiAs) through second step self-assembly

A 245nm

B 445nm

Size Ratio: 0.551

Stoichiometry: 1

Only A-B	A	B
A		+
B	+	+

Perspective view of sheared NaCl

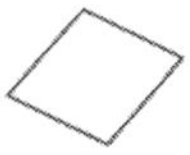

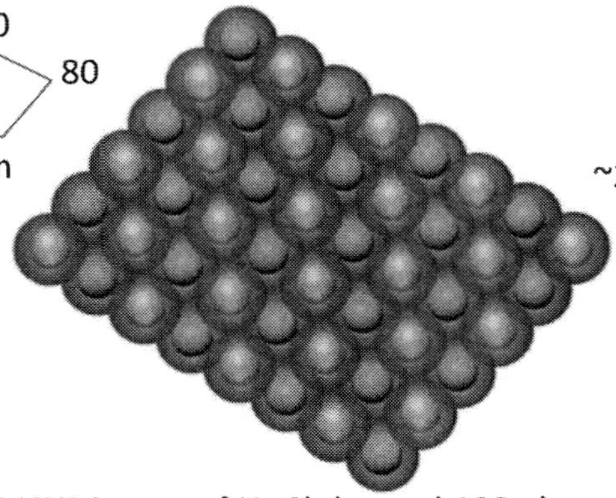

MAYA image of NaCl sheared 100 plane

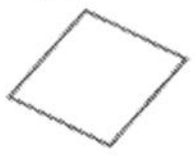

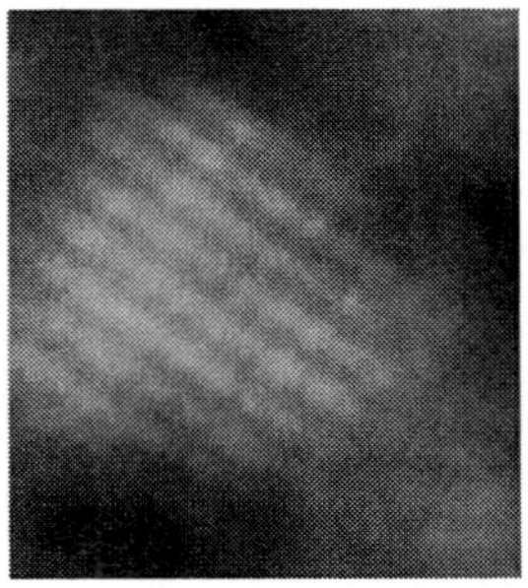

Confocal Image of NaCl sheared 100 Plane

Transformation Paths

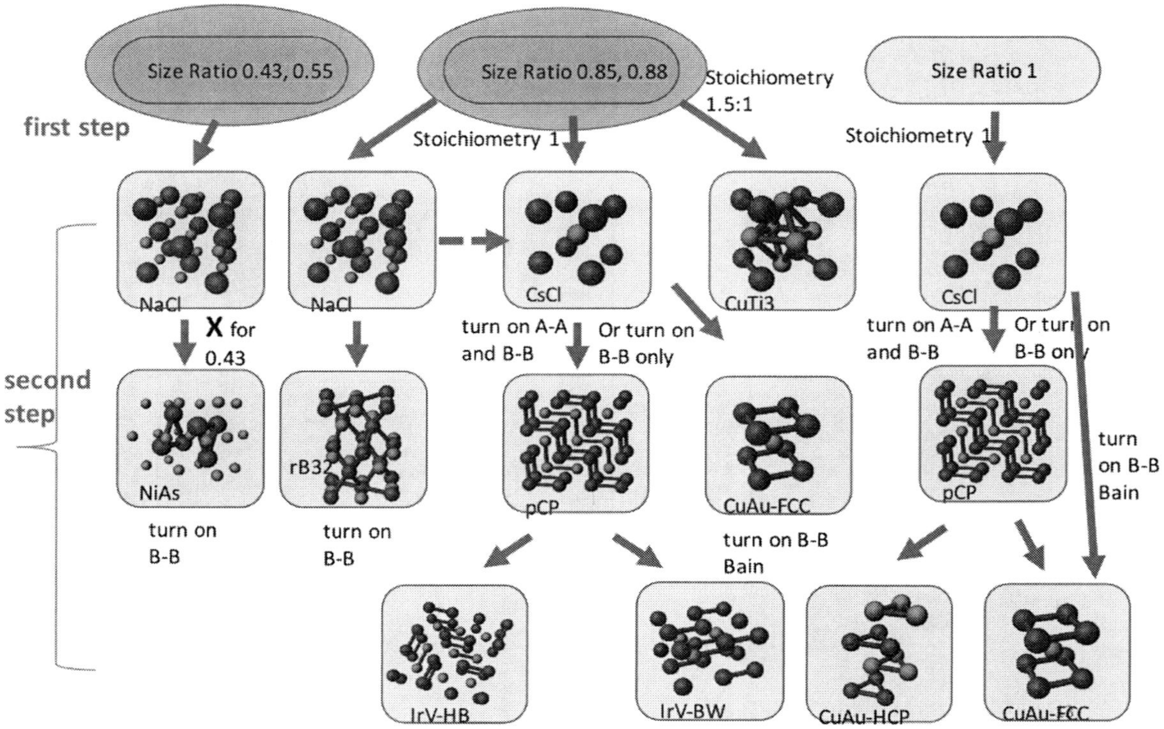

IrV Reconstruction

A 378nm

B 445nm

Size Ratio: 0.849

Stoichiometry: 1

A-A B-B A-B	A	B
A	+	+
B	+	+

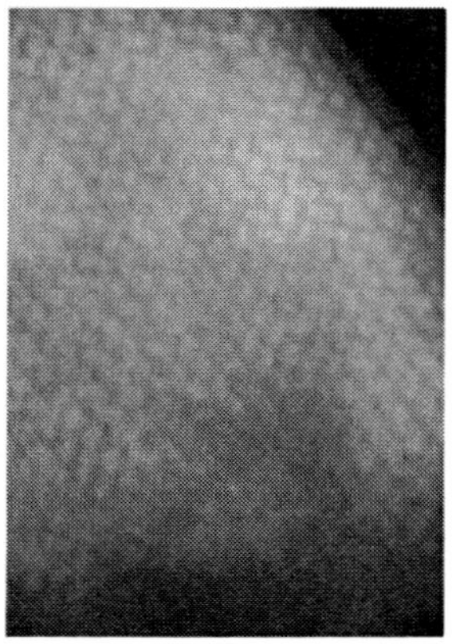

IrV_herringbone

IrV_Basket weave

Stacking Difference 14

Pseudo Close Pack (PCP) Structure as an intermediate

Credit to Ian and Talid

A 392nm

B 445nm

Size Ratio: 0.88

Stoichiometry: 1

A-A B-B A-B	A	B
A		+
B	+	+

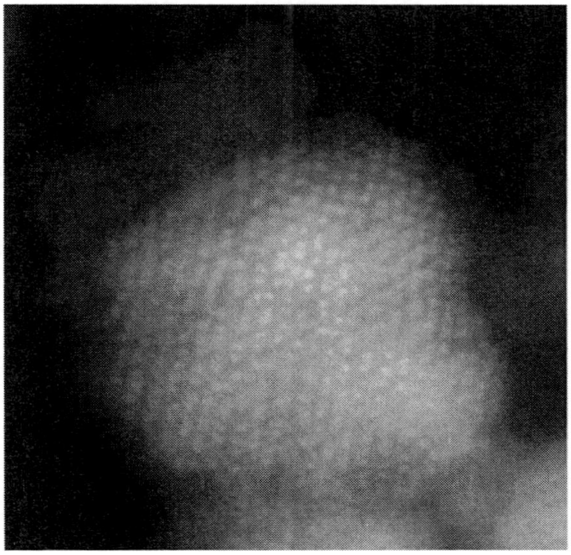

Confocal of PCP structure at 0.88 size ratio, green channel

Double Diamond (rB32)

A 392nm

B 445nm

Size Ratio: 0.88

Stoichiometry: 1

B-B A-B	A	B
A		+
B	+	+

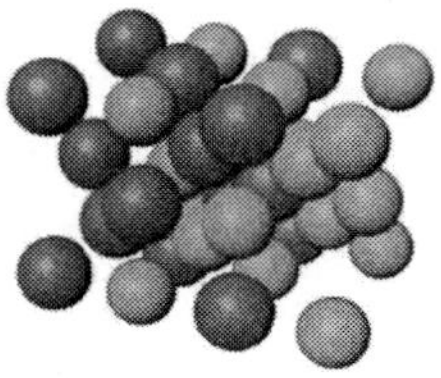

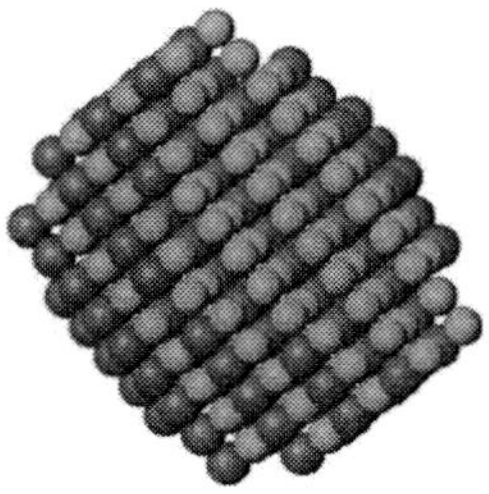

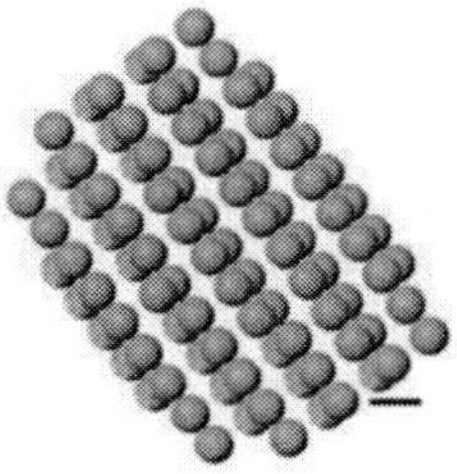

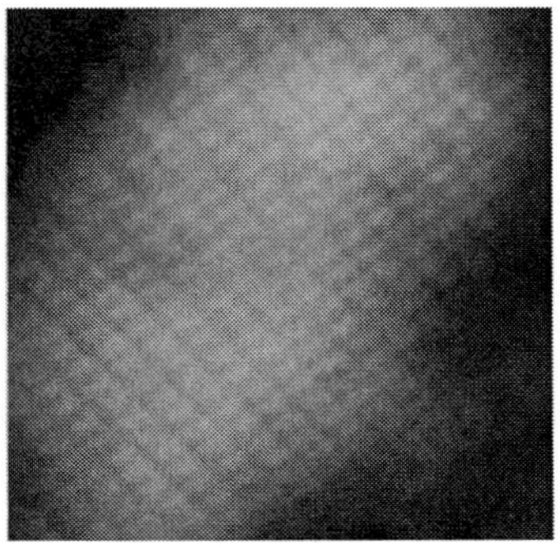

Double Diamond (211) plane

Double Diamond Crystal Proof

While the mechanism of the Double Diamond formation is still yet to study(rB32)
We think it came from NaCl

Colloidal crystals with diamond symmetry
at optical lengthscales, Yifan Wang et al., Nature Communication, 2017

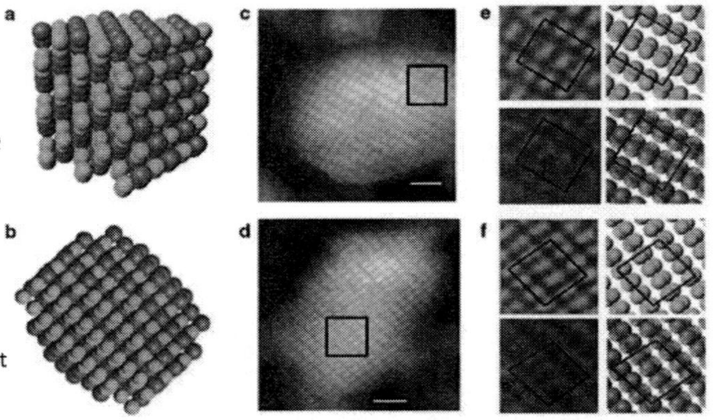

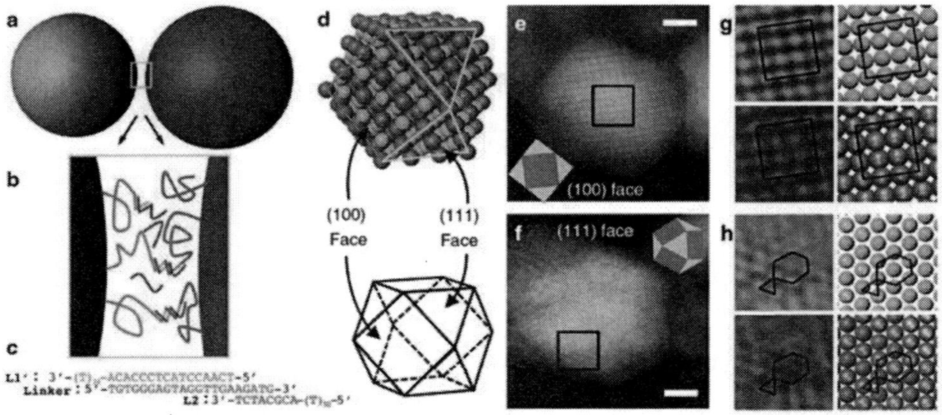

17

Transformation Paths

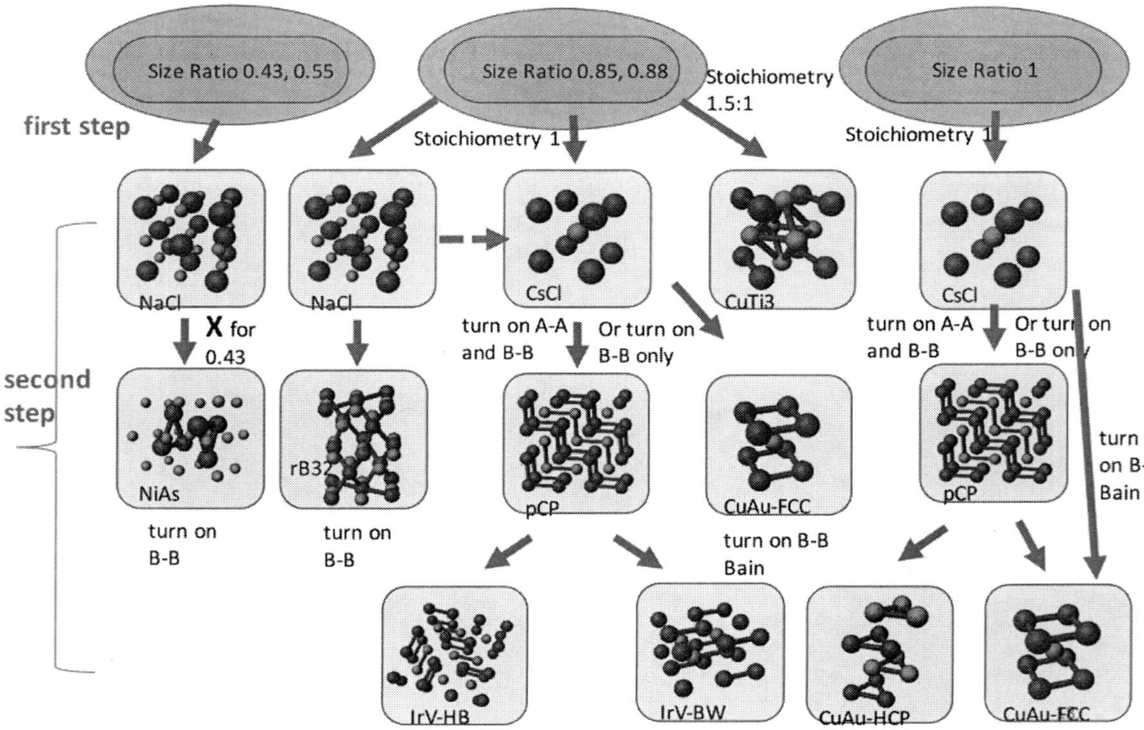

Conclusion and Future Plans

Conclusion:

1) We obtained DNA colloidal crystal from DNA-grafted spheres with high DNA concentration of different size ratios, imaged them and measured crystallography to determine the structure.
2) We applied different interaction matrix and α value studies to different size ratio systems and found the properties of each.
3) We proposed and further demonstrated different parent structures and their transformation mechanisms and paths.

Future Plans:

1) Keep on trying different parameters to build a library of DNA colloidal crystals made from various size ratios and shapes.
2) Apply on different interaction matrix and study on possible transformations.
3) Find ways to make different building blocks through new chemistry and DNA design and make 2nd order crystals through DNA cluster assembly.

19

Acknowledgement

Professor John Crocker

Doctor James McGinley

Doctor Taesoup Shim

Other Crocker lab Members:
Jennifer Hwang
Chris Porter
Mehdi Molaei
Ehsan Atefi

Thank you all for your attention and patience.

Any Questions?

yifanw@seas.upenn.edu

20

500871 Material Design By DNA-Mediated Interactions between Colloids

Abstract:

The ability to rationally design and synthesize complex nano-, micro-, and hierarchically structured materials with tunable functionality represents a holy grail that holds exciting promise for revolutionizing materials applications spanning catalysis, molecular sensing, drug delivery, molecular and charge manipulation, and beyond. Such applications hinge upon precise control over morphology, topology, material connectivity, and function, and demand accommodation of a wide range of materials compositions spanning organic, inorganic, and hybrid structures. One of the most promising approach to achieve such unprecedented control over colloidal self-assembly is to use DNA-mediated interactions between particles functionalized with partially complementary DNA sequences. In general, DNA functionalized particles have several adjustable parameters: the hybridization interaction strength, temperature, solution stoichiometry and DNA grafting density. Using molecular dynamics simulations of a coarse-grained model developed specifically to study such systems, we show that variations in these key factors provide selectivity between different lattice structures such as body-centered cubic (BCC) and close-packed (FCC and HCP) structures. In addition, we are able to program structures that interconvert via diffusionless transformations via external stimuli such as temperature changes.

485046 High Throughput Acoustically Driven Self-Assembly of Microfluidic Colloidal Crystals

Abstract:

Manufacturing defect-free ordered materials is immensely important for producing phononic and photonic crystals for energy applications or to serve as waveguides for optical computing. External fields (e.g., electric, magnetic, flow) that drive self-assembly are a form of bottom-up manufacturing, where micro- or nanoscale building blocks assemble spontaneously to form complex shapes. Drawbacks to current approaches include low throughput (~10 particles per second or less), the formation of permanent defects (e.g., grain boundaries, point defects, line defects) during assembly and the high cost associated with fabricating devices to produce external fields. In this presentation, we demonstrate a high throughput (~300 particles per second) self-assembly process based on acoustically-driven assembly of microparticles using of-the-shelf components without the need to access a clean room.

Fifteen micron polystyrene particles dispersed in water are continuously introduced to glass-capillary microfluidic flow cell with a syringe pump. A piezoelectric element attached to the size of the flow cell is driven to generate an acoustic standing wave that drives particles to assemble at acoustic nodes. An experimental state diagram shows the effect that fluid flow rate and applied acoustic pressure have the assembled microstructure. Highly ordered, continuous crystals are observed to assemble in less than a minute with a throughput rate of several hundred particles per second were observed under various pressure and flow rates. Particle tracking software is used to identify the location of microparticles assembled in ordered structures. This data is used to as part of a distribution analysis to understand the compressive influence that the acoustic pressure has on the observed structures. The flow cell throughput is quantified using a micro-particle image velocimetry analysis of optical video microscopy data. The degree of order within the assembled structures is quantified by comparing the number of nearest neighbors to theoretical expectations for an ideal ordered structure.

Future work is focused on mitigating defect formation in our ordered crystal structures. Acoustic assembly is dynamically controllable, allowing us to understand the kinetic effects that fluid flow and acoustic pressure have on dislocation dynamics. We aim to develop real-time annealing techniques to remove defects during crystal formation.

499771 The Role of Interaction Heterogeneity in Colloidal Crystallization

Abstract:

Heterogeneity in a population of particles is generally considered to be a detrimental factor for the self-assembly of ordered structures. While this is well-established for certain types of heterogeneity such as size polydispersity, here we show, using a combination of equilibrium and non-equilibrium simulations, that heterogeneity in the pairwise interaction strength among a collection of particles may in fact be useful for nucleation of crystalline phases. In particular, we consider interaction heterogeneities that may arise, for example, from particle-to-particle variations in the density of DNA oligomer brushes grafted on the surface of micron-scale spherical particles. In this setting, it is now well-known that the multivalent nature of DNA-mediated particle binding introduces constraints for assembly by increasing the temperature dependence of the resultant interparticle potential and consequently narrowing the operating window for high quality crystallization. For a given areal density of DNA strands, this effect becomes particularly pronounced for larger particles and represents a key challenge for DNA-driven assembly of micron-scale particles.

The beneficial impact of population-wide interaction heterogeneity is shown to arise from a synergistic combination of two effects. First, we employ equilibrium umbrella sampling simulations to show that heterogeneity lowers the free energy barrier associated with the nucleation of crystals by the formation of strongly-bound small clusters, even if the overall average binding energy between pairs of particles is held constant. This finding is confirmed by direct, non-equilibrium crystal growth simulations, which demonstrate that crystallization proceeds over a wider range of average interaction energy. On the other hand, these simulations also show that variations in the interaction strength between particles inhibit gelation and polycrystallinity by keeping the number of stable nuclei low, allowing individual nuclei to grow unhindered. In combination, these effects lead to a â€˜spreading-outâ€™ of the nucleation process and a concomitant widening of the crystallization window. These results are used to investigate potential avenues for intentionally controlling the nucleation rate in colloidal crystallization experiments.

500901 Self-Assembly of Open Structures Using Depletion

Abstract:

Depletion is a universal mechanism to induce order in colloidal self-assembly. In the absence of explicitly attractive forces or patchy interactions, anisotropic colloids usually crystallize into close-packed structures. Open structures are of high interest for applications for example as photonic crystals. Here we investigate the depletion-driven self-assembly of faceted colloids using Monte Carlo simulations. We find that such colloids can form open structures purely from hard shape and depletant entropy, as well as complex, close-packed unit cells. Our highly efficient simulation algorithm can readily be used to study the self-assembly of (almost) arbitrary convex and concave shapes with depletion.

499296 Controlled Self-Assembly of Colloidal Discotic Liquid Crystals

Abstract:

Colloidal anisotropic platelets are model systems for discotic liquid crystal investigation. We have systematically studied the ZrP monolayers due to the capability to control the size and surface chemistry. I will discuss our recent research on liquid crystal phase formation in colloidal discotic suspensions, particularly the controlled self-assembly under external fields, such as gravity, electric and magnetic field, and temperature gradient.

487724 Stratification Dynamics in Drying Colloidal Mixtures

Abstract:

Drying colloidal mixtures can undergo a spontaneous, nonequilibrium vertical separation (stratification) into layers enriched in particles of a specific size, an effect which can be exploited to create functional coatings. It was recently observed in experiments and simulations that for large particle size ratios and high drying rates, smaller colloids migrate to the top of the film, while big colloids are pushed to the bottom, creating a peculiar and counterintuitive â €œinvertedâ€ stratification. To investigate this behavior, we performed implicit-solvent molecular dynamics simulations of binary colloidal mixtures for a wide range of particle size ratios and drying rates, and quantified the stratification dynamics in detail. We found that the stratified layer of small colloids grew faster and to larger thicknesses for larger size ratios. Interestingly, inverted stratification was observed even at moderate drying rates, but the thickness of the stratified layer decreased. We propose a model based on dynamical density functional theory which captures the salient features of our simulations and can help to explain the observed phenomena.

493106 Theoretical and Experimental Investigation of Phase Separation in Noble Metal Nanoparticle Monolayers

Abstract:

Controlling the interfacial area of nanoparticles is a crucial component to controlling their properties due to their high surface area to volume ratio. One of the most common ways to control this interfacial area is with the use of a monolayer. Therefore, careful design of nanoparticle monolayers allows for a wide range of control of nanoparticle properties. Our work looks at the design rules that influence the morphology of a monolayer. To do this we use a combination of experimental and computational techniques to explore the effect of physical and chemical changes in the nanoparticle monolayer systems. Experimentally we use Transmission Electron Microscopy (TEM) to assess nanoparticle shape and size and Matrix-Assisted Laser Desorption and Ionization (MALDI) to assess the degree of order present in nanoparticle monolayers. In addition, we model the systems using both Self-Consistent Field Theory (SCFT) and atomistic simulations to model these nanoparticle monolayers. We are then able to calculate predicted MALDI spectrum from these simulations which allows us to directly compare theory and experiment. This direct comparison allows us to have greater confidence in the insights provided by our computational simulation.

499798 Bending-Induced Buckling Instabilities in Self-Assembled Elastomeric Composite Films

Abstract:

Buckling instabilities induced by various mechanical stresses are universal phenomena observed at all length scales in a wide range of materials (e.g., polymers, metals, and ceramics) in both natural and manmade systems. Bilayer buckling systems consisting of rigid skin films attached to thick compliant polymer foundations have been extensively investigated. Unfortunately, it is challenging to create a universal bilayer buckling platform that can cover a broad range of length scales, while simultaneously can achieve high structural stability and system durability. Additionally, bending-induced buckling instabilities of bilayer composite systems possessing discrete rigid phases, which are of great technological and scientific importance in developing next-generation wearable and flexible devices and new non-linear instability models, are far from being well studied. Here we report a new and universal bilayer buckling system by applying scientific principles drawn from two disparate fields that do not typically intersect â€“ the mature polymer buckling and colloidal self-assembly techniques. Self-assembled colloidal crystals embedded in elastomers provide a model buckling system exhibiting many unique properties, such as high effective elastic moduli of the composite skin layers, well-controlled skin-layer thicknesses spanning at least two orders of magnitude (from 100 nm to 10 Âµm), unconventional buckling microstructures comprising discrete rigid phases dispersed in a continuous elastic matrix, and high mechanical durability. The fundamental mechanisms of bending-induced buckling instabilities exhibited by the self-assembled elastomeric composite films have been investigated by mechanical finite element analysis. The numerical simulations match well with the experimental results.

501561 Molecular Modeling of Microstructure, Solubilization and Micro-Emulsion of Block Copolymer Micelles By iSAFT

Abstract:

Block copolymer micelles have been the focus of much interest for both academia and industry over decades for extensive applications in enhanced oil recovery, cosmetic products, and drug delivery. The immiscibility of constituents causes micro-phase separation for high density block copolymer melts, and micellar aggregates when it is dissolved in a selective solvent. A few chemical potential models have been developed to predict sizes, aggregation number, solubilization ability, critical micelle concentration(CMC) etc. Most of the theories are phenomenological equation of state(EoS) types of theories and thus cannot fully capture the physics at a molecular level. We study the microstructures using inhomogeneous statistical associating fluid theory(iSAFT) a classical density functional theory iSAFT takes into account a range molecular interactions, including hard-sphere repulsion, chain connectivity, van der Waals attraction, and association.

iSAFT is demonstrated to predict different influences of hydrophobic tail and hydrophilic head of block copolymer surfactants on the size and CMCs of micelles. iSAFT showed that radii and CMCs of micelles are strongly dependent on the length of hydrophobic tail but not for the length of hydrophilic head. This agrees with experimental finding of Pluronic micelles. iSAFT also predicts the radii of triblock surfactant micelles is smaller than of diblock copolymer surfactants with the same head and tail groups. Similar agreement between theory and experiment results is also found for the CMCs of micelles of triblock and diblock copolymer surfactants.

Structure of solubilized micelles is similar to a surfactant micro-emulsion. By predicting the distribution of solute, solvent and surfactants within micelles by iSAFT, we are able to show the enhanced solubility of hydrocarbon in water with surfactants. iSAFT predicts the different solubility of normal alkanes and branched alkanes with different carbon numbers when using different surfactants.

AUTHOR INDEX

AUTHOR INDEX